INVITATION TO GEOMETRY

Z. A. MELZAK
Department of Mathematics
University of British Columbia

DOVER PUBLICATIONS, INC.
Mineola, New York

Bibliographical Note

This Dover edition, first published in 2008, is an unabridged republication of
the work originally published in 1983 by John Wiley & Sons, Inc., New York.

Library of Congress Cataloging-in-Publication Data

Melzak, Z. A., 1926–
    Invitation to geometry / Z. A. Melzak. — Dover ed.
      p. cm.
    Originally published: New York : Wiley, 1983.
    Includes bibliographical references and index.
    ISBN-13: 978-0-486-46626-2
    ISBN-10: 0-486-46626-4
      1. Geometry. I. Title.

QA445.M43 2008
516—dc22

                                                            2007051604

Manufactured in the United States of America
Dover Publications, Inc., 31 East 2nd Street, Mineola, N.Y. 11501

PREFACE

This book grew out of the notes assembled for a full-year course in geometry that I have given several times at my university. The course was intended for students with only a modest background in mathematics beyond the high-school level, say, with some knowledge of calculus and of linear algebra. Many of them were prospective schoolteachers; others came from computer science, commerce, architecture, geography, arts, and even physical education or the premedical and prelaw groups. Such students often need a certain number of credits in mathematics to get their degrees, yet the offerings for them at most North American universities are skimpy. Certainly geometry is the best subject here, with elementary number theory as the only serious competitor, yet it is sadly neglected. For many of my students my course was what is hideously though correctly described as the terminal mathematics course. This put a heavy premium on making it as exciting as the circumstances allowed, especially for prospective teachers so that they would have some material and tools to interest *their* students, but of course for others as well.

This book is so arranged that it could be adapted for several types of geometry course, of full-year or one-semester duration, or for collateral reading in other subjects such as calculus, optimization, operations research, and so on. The material is handled in self-contained and independent chapters that present introductions to several branches of geometry: classical Euclidean material, polygonal and circle isoperimetry, conics and Pascal's theorem, geometrical optimization, geometry and trigonometry on a sphere, graphs, convexity, elements of differential geometry of curves.

Additional material may be conveniently introduced in several places. For instance, in Chapter 4 certain symmetry properties are considered that go with the cube or the torus; here it is possible to bring in more algebra centering on symmetry and groups. The winding of curves on the torus is briefly handled in Chapter 4, and certain elementary connectivity properties in Chapter 9; in these places it is possible to add some further simple descriptive topology. The extension of Euclidean to the projective plane appears in Chapter 2 and provides a convenient place for some projective geometry.

The choice of topics and the treatment of those that were selected depended on several factors. In the first place, as mentioned before, there was the desire to make the material interesting. A rough criterion here was something like this: Exclude anything that is likely to leave the students' memories as soon as they leave the examination room. In the second place, I have chosen certain things that seemed useful or, to use a suspect word, relevant. Considerable effort has gone into emphasizing connections with and illustrations from such subjects as calculus and simple mechanics. In more than one place the principal aim has been to develop the students' grasp of spatial relationships. This was done in the belief that the modern superabundance of paper, book pages, screens, blackboards, and other surfaces has rendered us in effect two-dimensional in our thinking and visualizing. Certain rather important topics, such as analytic geometry, geometrical transformations, use of vectors in geometry, and area and volume measurement, are omitted since they are adequately covered in existing texts and, especially, since most good calculus books devote some space to them.

The entire book can be comfortably covered in a full-year course. In the matter of choosing what to teach for a single-semester course, I am certain that the instructors can do their own selecting very well to suit the circumstances, but a few suggestions on this follow.

Type of one-semester course	Chapters suggested
Basic geometrical core	1, 4, 5, 6, 10
Slant on pure geometry	1, 2, 5, 8, 9, 10
Slant on applications	1, 3, 4, 6, 7, 8
For collateral work, e.g., to calculus	1, 3, 4, 5, 6, 8
For more historical background	1, 3, 5, 7, 8

Each chapter ends with some exercises of varying degrees of difficulty, and there is at the end of the book a small bibliography for further, collateral, and background reading. Some informal notes are also given at the end, providing a sort of running commentary on the material of the text. The benefit of help from Dr. Margaret Rayner, of St. Hilda's College, Oxford, and from Dr. Armin Frei and the late Dr. Ronald Riddell, both of the University of British Columbia, is gratefully acknowledged. They have read and criticized parts of the text and supplied a number of useful suggestions.

Z. A. MELZAK

Vancouver, British Columbia
November 1982

CONTENTS

INVITATION TO GEOMETRY

CHAPTER ONE

HERON'S FORMULA AND RELATED ONES

The area S of the triangle of sides a, b, c is given by the formula

$$S = \sqrt{s(s - a)(s - b)(s - c)}, \qquad s = \frac{a + b + c}{2}, \tag{1}$$

traditionally ascribed to Heron of Alexandria, a Greek mathematician of the first century A.D. To prove it introduce the auxiliary quantities x, y, z of Fig. 1.1a, getting the equations

$$x + y = c, \qquad x^2 + z^2 = b^2, \qquad y^2 + z^2 = a^2.$$

Subtract the last two equations and divide the result by the first one, obtaining

$$x - y = \frac{b^2 - a^2}{c}.$$

Now $x + y$ and $x - y$ are known; solving for x

$$x = \frac{b^2 - a^2 + c^2}{2c}, \tag{2}$$

and since $z^2 = b^2 - x^2$

$$z^2 = \frac{4b^2c^2 - (b^2 - a^2 + c^2)^2}{4c^2}. \tag{3}$$

At this point we might ask: What if the configuration is as in Fig. 1.1b, with an obtuse angle? Here we have

$$x - y = c, \qquad x + y = \frac{b^2 - a^2}{c},$$

instead of

$$x + y = c \qquad x - y = \frac{b^2 - a^2}{c},$$

1

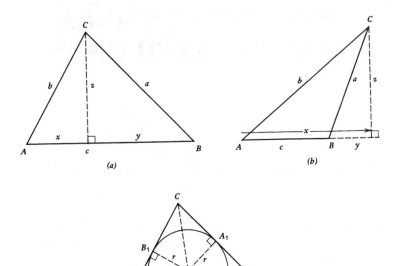

(a)   (b)

(c)

Figure 1.1

as before. Therefore x and z are still given by (2) and (3) even though x + y and $x - y$ have interchanged their values. Since the area S is $cz/2$, we find from (3) that

$$16S^2 = 4b^2c^2 - (b^2 - a^2 + c^2)^2. \qquad (4)$$

Recall now the basic factorization formula

$$u^2 - v^2 = (u + v)(u - v).$$

The R.H.S. of (4) is a difference of squares; therefore

$$16S^2 = (2bc + b^2 - a^2 + c^2)(2bc - b^2 + a^2 - c^2),$$

which may be written as

$$16S^2 = [(b + c)^2 - a^2][a^2 - (b - c)^2].$$

Each factor is again a difference of squares; factorizing these we obtain

$$16S^2 = (a + b + c)(-a + b + c)(a + b - c)(a - b + c).$$

For convenience, the semiperimeter $s = (a + b + c)/2$ is introduced; the four preceding factors then become $2s$, $2s - 2a$, $2s - 2c$, $2s - 2b$, and we get (1).

Why do we need formula (1)? We note its pleasing appearance as well as its natural symmetry: If $S = S(a, b, c)$ is considered as a function of a, b, c, then S remains unchanged by any permutation of its variables:

$$S(a, b, c) = S(a, c, b) = S(b, a, c) = S(b, c, a) = S(c, a, b) = S(c, b, a).$$

This is just as it should be for the area of the triangle in terms of its sides. However, this symmetry under permutations is not at all obvious in formula (4), which also gives S in terms of a, b, c.

What is perhaps more important, formula (1) allows a convenient computation of S by logarithms: After the additions and subtractions are carried out, all the multiplications as well as the extraction of the square root can be done by means of logarithms. Thus esthetics and utility combine here even though it must be remembered that (1) precedes logarithms by some 1500 years.

The arithmetic convenience of computing by logarithms carries over from (1) to some trigonometric calculations. Suppose, for instance, that the sides a, b, c of a triangle are known and its angles A, B, C are to be found. Note first the cosine law

$$a^2 = b^2 + c^2 - 2bc \cos A \tag{5}$$

that results from (2) on putting $x = b \cos A$ (see Fig. 1.1a). It follows from (5) that

$$\cos A = \frac{b^2 - a^2 + c^2}{2bc},$$

but this again is not suited to logarithmic computation. To obtain what is needed we reason as follows: Let O be the center of the circle inscribed into our triangle, and let r be its radius, as shown in Fig. 1.1c. Recall that O is the point at which the angle bisectors of the triangle intersect. We have then

$$\Delta ABC = \Delta ABO + \Delta BCO + \Delta CAO$$

and expressing the areas of these triangles we get $S = rs$. Hence, as a bonus, comes the simple expression for the in-circle radius r: $r = S/s$ or using (1)

$$r = \sqrt{\frac{(s - a)(s - b)(s - c)}{s}}.$$

AC_1 can be expressed in terms of a, b, c, s. Note first that

$$BA_1 + A_1C = a, \qquad AC_1 + BA_1 + A_1C = \tfrac{1}{2}(AB + BC + CA) = s. \qquad (6)$$

We observe that

$$AC_1 + BA_1 + A_1C = AC_1 + BA_1 + CB_1$$

and the last sum contains just one segment from the three pairs of them, making up the circumference of the triangle. By subtraction in (6) $AC_1 = s - a$ and by reference to Fig. 1.1c $\tan A/2 = r/AC_1$ so that

$$\tan \frac{A}{2} = \sqrt{\frac{(s - b)(s - c)}{s(s - a)}}$$

with analogous expressions for $\tan B/2$ and $\tan C/2$, all of them adapted to logarithmic computation.

Heron's formula generalizes to quadrilaterals even though here we cannot expect a formula as simple as (1). Part of the reason for it is that the four sides a, b, c, d alone do not determine a quadrilateral, whereas the three sides a, b, c do determine the triangle. Consider the quadrilateral Q of Fig. 1.2a, with the sides a, b, c, d and the diagonals x, y. We compute y twice by the cosine law, once from $\triangle ABD$ and once from $\triangle BCD$, getting

$$y^2 = a^2 + d^2 - 2ad \cos A = b^2 + c^2 - 2bc \cos C.$$

By subtraction it follows that

$$ad \cos A - bc \cos C = \frac{a^2 + d^2 - b^2 - c^2}{2}. \qquad (7)$$

Obtaining this expression illustrates the frequently useful technique of computing a quantity in two ways and equating the results. Next, the area S of Q is the sum

$$\text{area } \triangle ABD + \text{area } \triangle BCD.$$

Since the area of a triangle is half the product of any two sides times the sine of the angle formed by those sides, we have

$$ad \sin A + bc \sin C = 2S. \qquad (8)$$

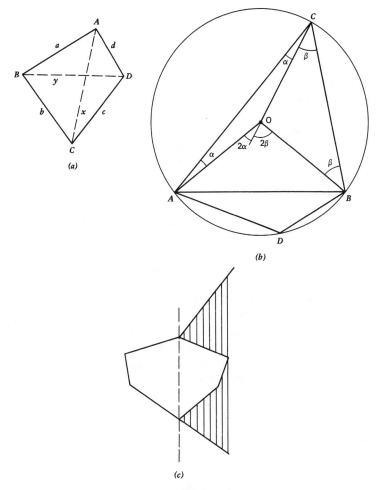

Figure 1.2

Now (7) and (8) are squared and added. Why do this? One answer is: because it works. But considerable simplification may be expected in advance since $\sin^2 A + \cos^2 A = 1$. We get thus

$$16S^2 + 16abcd\cos^2\alpha = 4(ad + bc)^2 - (a^2 + d^2 - b^2 - c^2)^2, \quad (9)$$

where $\alpha = (A + C)/2$. In obtaining (9) we have used the trigonometric formulas

$$\cos (A + C) = \cos A \cos C - \sin A \sin C,$$

$$\cos 2\alpha = 2 \cos^2\alpha - 1.$$

Incidentally, if we chose to work with the other diagonal, that is, x instead of y, then we would have had $\alpha = (B + D)/2$ instead of $\alpha = (A + C)/2$. However, since $A + B + C + D = 360°$ the value of $\cos^2\alpha$ is the same, either way.

Formula (9) is processed in the same way as (4) before. The R.H.S. is a difference of two squares and each factor is again a difference of two squares, giving us

$$S^2 = (s - a)(s - b)(s - c)(s - d) - abcd \cos^2\alpha, \qquad (10)$$

where s plays the same role as before: It is the semiperimeter $(a + b + c + d)/2$ of the quadrilateral Q. Formula (10) is the quadrilateral counterpart of Heron's formula (1), which is a special case: If $d = 0$ in (10), the quadrilateral Q reduces to a triangle and (10) becomes (1).

In particular let $\alpha = \pi/2$ so that $\cos \alpha = 0$. We then have $A + C = 180°$, which means that our quadrilateral is cyclic: Its four vertices lie on a circle. This follows from the theorem of Thales, which states that all angles in a circle, on a given chord, are equal. The theorem is illustrated in Fig. 1.2b. O is the center of the circle and we find that for any position of C on the circle $\sphericalangle BCA = \frac{1}{2}\sphericalangle BOA$. If D is any point on the circle but on the other side of the chord AB from C, then $\sphericalangle BDA = \frac{1}{2}\sphericalangle$ (concave) BOA so that

$$\sphericalangle BCA + \sphericalangle BDA = \tfrac{1}{2} \cdot 360° = 180°.$$

Here the concave angle is the one that exceeds 180°. Conversely, if C and D are such that the preceding angle condition holds, then the points A, B, C, D lie on a circle. We can express it thus: A necessary and sufficient condition for a quadrilateral to be cyclic is that opposite angles add up to 180°.

When $\alpha = \pi/2$, (10) reduces to

$$S = \sqrt{(s - a)(s - b)(s - c)(s - d)}, \qquad s = \frac{a + b + c + d}{2}. \qquad (11)$$

This formula for the area of a cyclic quadrilateral is due to the Hindu mathematician Brahmagupta of the sixth century A.D. If only the sides a, b, c, d of Q are given, then it follows from (10) that the quadrilateral

Q has the largest area when it is cyclic. That is, of all quadrilaterals with four given sides the one inscribed into a circle has the largest area.

We shall extend this proposition to an arbitrary polygon P with n sides, $n \geq 4$, of lengths $a_1, a_2, \ldots, a_n$, in order. Let these sides be fixed but not the angles between them. When is the area S of P greatest? It is clear that P must be convex; otherwise we could reflect the reentrant parts outward, as is shown in Fig. 1.2c, and thereby increase the area S while keeping all the sides of the same lengths. Now consider in P any three consecutive sides together with that diagonal of P that completes those three sides to a quadrilateral Q. Keeping the rest of P rigid and varying Q only, we note that the area of Q, and hence of P, is greatest when Q is cyclic. However, the same argument applies to any three consecutive sides. We conclude that if the area of P is greatest, then any four consecutive vertices of P lie on a circle. But three consecutive vertices already determine a unique circle; hence all such four-vertex circles must coincide. In other words, P has the greatest area when it is cyclic, that is, when all its vertices lie on a circle.

Next, consider the same problem of maximizing the area of an n-gon P but now allow the sides of P to vary in length supposing only that their number n and their total length $a_1 + a_2 + \cdots + a_n$ stay fixed. As was shown before, P must be cyclic. But, furthermore, any two consecutive sides are of equal length. This is proved by fixing all the vertices $A_1, A_2, \ldots, A_n$ of P except one of them, say, A_i. Consider the triangle $T = A_{i-1}A_iA_{i+1}$, let $A_{i-1}A_{i+1} = 2a$, and let A_i vary so that $A_{i-1}A_i + A_iA_{i+1}$ stays fixed. We then have $A_{i-1}A_i = b - x$, $A_iA_{i+1} = b + x$. The circumference of P stays fixed as does that of T. When is the area greatest? Only the area S of T need be considered, and by Heron's formula (1)

$$S^2 = (b + a)(b - a)(a + x)(a - x) = (b^2 - a^2)(a^2 - x^2).$$

Hence S is greatest when $x = 0$, that is, when $A_{i-1}A_i = A_iA_{i+1}$. Since i is arbitrary, any two consecutive sides are of equal length; hence all of them are equal. This, together with the cyclicity of P, shows that among all n-gons of fixed total perimeter, the regular n-gon encloses the greatest area.

As an important consequence we obtain the isoperimetric theorem: Of all plane closed curves C of fixed circumference the circle encloses the greatest area. First, C must be convex, for otherwise we could use the outward reflexion technique as for the polygons. This reflexion preserves the total length of C while increasing the area enclosed. Next, we can approximate to C arbitrarily well by an n-gon, with n sufficiently large. For instance, we might take an inscribed n-gon with equal sides. Since every such approximating n-gon must be regular, it follows that C is the circle.

Of course, to make the preceding argument rigorous it would be necessary to define the length of a curve precisely and to employ the ϵ-δ techniques for limits. Among other things it would be necessary to show that a maximizing curve C exists. This may appear to be only a subtlety, but such an existence proof is necessary for full rigor. The danger of assuming that a solution exists is well illustrated by the following statement: Of all positive integers 1 is the greatest because all others are increased by squaring. Here the mistake lies solely in assuming that the greatest positive integer exists, and the correct conclusion is: If there exists a greatest positive integer, then it is 1 because all others are increased by squaring.

It may have occurred to us that in the preceding arguments we have assumed the existence of cyclic n-gons with prescribed sides and that this really requires a proof. We therefore prove the following realizability theorem: Let a_1, a_2, . . . , a_n be n positive numbers, $n \geq 3$; then there exists a convex n-gon P with sides of lengths a_1, a_2, . . . , a_n, in this order, if and only if each a_j is less than the sum of all the other a's:

$$a_j < \sum_{\substack{i=1 \\ i \neq j}}^{n} a_i, \qquad j = 1, 2, \ldots, n. \qquad (12)$$

If this condition is satisfied, then P can be taken to be cyclic. This cyclic n-gon is unique.

We start by observing that the condition (12) may be rewritten: If a_M is the largest number among the a's, then

$$a_M < a_1 + \cdots + a_{M-1} + a_{M+1} + \cdots + a_n \qquad (13)$$

for (12) and (13) are equivalent. If it should happen that the largest value among the a's is assumed by several numbers, say $a_{M_1}, a_{M_2}, \ldots$, then any one index $M_1, M_2, \ldots$, but only that one, is used as M.

The necessity of (13) for the existence of our convex n-gon P is simple: If such P exists, then by projecting all other sides onto the line containing the longest side we find that (13) holds. To prove the sufficiency we shall show that if (13) holds, then a cyclic n-gon P can be found.

Suppose then that (13) holds and take a circle C of sufficiently large radius, for instance, of radius equal to $a_1 + \cdots + a_n$. Starting from a fixed initial point I on C we lay off consecutive chords in C of lengths a_1, $a_2, \ldots, a_n$; let F be the endpoint of the last one. Keeping the point I fixed, we decrease the radius of C so that all vertices of our polygonal line of chords, except for I, slide along the constricting circle. In particular, F is sliding along toward I.

There are now two possibilities. Either F can reach I so that the polygonal line closes up, and we are then finished, or at some time before F reaches I the process stops and we cannot decrease C any more because one of the chords has become the diameter of C. It is clear that this hold-up side is the one of maximum length a_M; let its endpoints be A and B. Now we reverse our process: Keeping A and B fixed we increase the radius of our circle C so that I and F again slide toward each other. If the ends I and F can be made to meet, we are again finished.

Suppose finally that this is impossible no matter how large the radius R of C becomes. Then it is easily shown, by taking R large enough so that all the sides except AB are almost collapsed onto AB, that condition (13) is not satisfied. This is a contradiction and the proof is complete.

The uniqueness of our cyclic n-gon is obvious from the construction. Further, the order of any two consecutive sides of a cyclic n-gon can be interchanged and the polygon remains a cyclic n-gon with the same circumscribing circle. It follows that all the different cyclic n-gons, with sides prescribed in length but not in their order, have the same circumscribing circle. Hence it follows also that they have the same area.

On the basis of the analogy with triangles let us call a cyclic n-gon P obtuse if the center of its circumscribing circle lies outside P, right if it lies on P, and acute if it lies inside P. The geometrical construction used in the proof of the realizability theorem enables us to derive the conditions for a cyclic n-gon to be obtuse, right, or acute. It is observed that the obtuse case occurs if and only if the maximal side a_M becomes the diameter of the circle C that is constricting. Since all the other sides then subtend angles at the center of C (i.e., the middle of a_M), which add up to less than 180°, we find that

$$\sum_{\substack{i=1 \\ i \neq M}}^{n} \arcsin \frac{a_i}{a_M} < \frac{\pi}{2}$$

is the necessary and sufficient condition for obtuseness. The general criterion is similarly obtained: Any cyclic n-gon constructed with the sides $a_1, a_2, \ldots, a_n$, in any order, is obtuse, right, or acute, depending on whether the sum

$$\sum_{\substack{i=1 \\ i \neq M}}^{n} \arcsin \frac{a_i}{a_M}$$

is less than, is equal to, or exceeds $\pi/2$. If R is the radius of the circumscribing circle, then we can show in the same way that

$$\sum_{i=1}^{n} \arcsin \frac{a_i}{2R} = \pi$$

in the acute case, and

$$\sum_{\substack{i=1 \\ i \neq M}}^{n} \arcsin \frac{a_i}{2R} = \arcsin \frac{a_M}{2R}$$

for the obtuse case. Either equation may be used for the right case, since then $R = a_M/2$. The last two equations, considered as equations for the unknown quantity R, determine R uniquely.

A simple variant of the constricting-circle construction can be used to prove other, similar propositions. For instance, it can be shown thus that any fixed polygon, convex or not, lies inside a circle that passes through at least three of its vertices. As in the proof of the realizability theorem, we exploit here a certain mechanical rigidity that goes with extremal geometric configurations and, dually, the slackness that goes with geometric configurations that are not extremal.

We turn now to the last topic of this section, Ptolemy's theorem. It is due to Claudius Ptolemaeus, an Alexandrian astronomer of the second century A.D. after whom the Ptolemaic system of astronomy, preceding that of Copernicus, is named. Ptolemy's theorem is as follows: Let $ABCD$ be any convex quadrilateral as shown in Fig. 1.3a; then

$$xy \leq ac + bd \tag{14}$$

and the equality holds if and only if the vertices A, B, C, D lie on a circle. Hence a convex quadrilateral is cyclic if and only if the product of its diagonals equals the sum of the products of opposite sides.

We remark that the assertion (14) of Ptolemy's theorem applies to nonconvex quadrilaterals as well. In such a case we reflect two sides by a mirror reflection in a diagonal, obtaining a convex quadrilateral in which

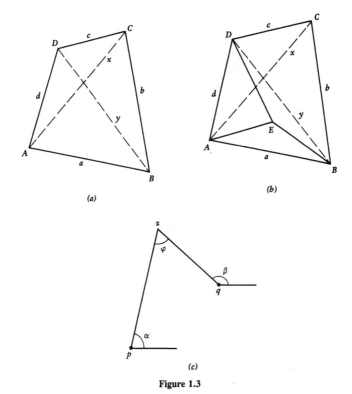

Figure 1.3

all sides are the same, one diagonal remains the same, and the other is increased.

To prove Ptolemy's theorem we use the auxiliary point E, shown in Fig. 1.3b, and determined so that $\triangle AED$ and $\triangle ABC$ are similar with A, E, D corresponding to A, B, C. It follows by proportionality that

$$\frac{AC}{AD} = \frac{BC}{DE} \quad \text{or} \quad x \cdot DE = bd. \qquad (15)$$

Also

$$\frac{AC}{AD} = \frac{AB}{AE} \quad \text{and} \quad \sphericalangle DAC = \sphericalangle EAB \qquad (16)$$

because

$$\sphericalangle DAC = \sphericalangle DAE - \sphericalangle CAE,$$

$$\sphericalangle EAB = \sphericalangle CAB - \sphericalangle CAE,$$

$$\sphericalangle DAE = \sphericalangle CAB.$$

It follows from (16) that ΔACD and ΔABE are similar with A, C, D corresponding to A, B, E. Hence

$$\frac{AC}{AB} = \frac{DC}{EB} \quad \text{or} \quad x \cdot EB = ac. \tag{17}$$

By adding (15) and (17) it follows that

$$x(DE + EB) = ac + bd.$$

This proves (14) since in ΔEBD we have

$$y = DB \leq DE + EB. \tag{18}$$

To finish the proof we have to find the necessary and sufficient condition for equality in (14) or, what is the same thing, in (18). This condition is that E lies on DB or, equivalently, that

$$\sphericalangle ADE = \sphericalangle ADB = \sphericalangle ACB.$$

But this means that AB makes the same angle at D and at C; hence the vertices A, B, C, D lie on a circle.

We sketch now an alternative proof based on the elementary properties of complex numbers. Let the complex numbers z_1, z_2, z_3, z_4 correspond to the vertices A, B, C, D. We verify the identity

$$(z_4 - z_2)(z_3 - z_1) = (z_4 - z_1)(z_3 - z_2) + (z_4 - z_3)(z_2 - z_1), \tag{19}$$

which holds for any four complex numbers. Next we recall the basic properties of complex numbers

$$|uv| = |u||v|, \qquad |u + v| \leq |u| + |v|;$$

applying these to (19) we get

$$|z_4 - z_2||z_3 - z_1| \leq |z_4 - z_1||z_3 - z_2| + |z_4 - z_3||z_2 - z_1|, \tag{20}$$

which is exactly (14). It remains to find out when the equality could occur in (14) or, equivalently, in (20). This comes down to the same thing as in our geometrical proof but instead of the triangle inequality

$$DB \le DE + EB \tag{21}$$

in its geometrical form, we use its complex-number version

$$|u + v| \le |u| + |v|. \tag{22}$$

The condition for equality in (22) is the same as for vectors: Equality holds if and only if u is a positive real multiple of v. In terms of (20) this condition is that the quotient

$$\frac{z_4 - z_1}{z_4 - z_3} \frac{z_2 - z_1}{z_2 - z_3} \quad \text{is a negative real number.} \tag{23}$$

Now, for complex numbers z, p, q as in Fig. 1.3c, we find that

$$\arg \frac{z - q}{z - p} = \arg(z - q) - \arg(z - p) = \beta - \alpha = \phi.$$

Since the argument of a negative real number is π, we find from (23) that the opposite angles, say ϕ and ψ, in our quadrilateral add up to π. This is precisely the condition for the quadrilateral to be cyclic, as was shown earlier.

We consider now several of the many consequences of Ptolemy's theorem. First, let $ABCD$ be the cyclic quadrilateral of Fig. 1.4a; the diagonal $AC = 1$ is a diameter of the circumcircle. We have then

$$AB = \sin \beta, \qquad BC = \cos \beta, \qquad CD = \cos \alpha,$$

$$AD = \sin \alpha, \qquad BD = \sin(\alpha + \beta).$$

Therefore, applying Ptolemy's theorem, we get

$$\sin(\alpha + \beta) = \sin \alpha \cos \beta + \cos \alpha \sin \beta,$$

which is the addition theorem for the sine function. As a matter of historical and practical interest, Ptolemy's theorem, in the preceding form of addition theorem for sine, was used to construct the early trigonometric tables.

Next, let ABC be an equilateral triangle and let P be any point on its circumcircle, for instance, on the shorter arc AB, as shown in Fig. 1.4b. Applying Ptolemy's theorem to the cyclic quadrilateral $APBC$, we find that

$$AP + BP = CP, \tag{24}$$

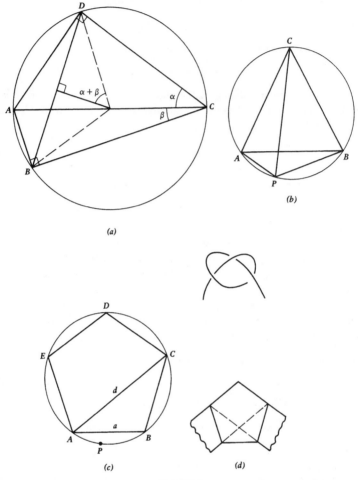

Figure 1.4

independent of the position of P on its circle. Continuing in the same fashion, we consider now a regular pentagon $ABCDE$ and a point P on its circumcircle, as shown in Fig. 1.4c. Let a be the side of the pentagon and d its diagonal. Then, applying Ptolemy's theorem to the cyclic quadrilaterals $ABCE$, $APBE$, $APBC$, $APBD$, we get

$$ad + a^2 = d^2,$$

$$d \cdot AP + a \cdot BP = a \cdot EP,$$

$$a \cdot AP + d \cdot BP = a \cdot PC, \tag{25}$$

$$d \cdot AP + d \cdot BP = a \cdot DP,$$

from which it follows by eliminating a and d that

$$AP + DP + BP = EP + CP \tag{26}$$

independent of the position of P. Further, choose P to bisect the arc AB, so that $AP = BP$ is the side x of the regular decagon inscribed into the circle, and let the circle radius be r. Then applying Ptolemy's theorem to the cyclic quadrilateral $APBD$ we find that

$$ar = xd.$$

Together with the first equation of (25) this gives us

$$\frac{r}{x} = \frac{d}{a} = \frac{a + d}{d} = \frac{1 + d/a}{d/a}.$$

Therefore the ratio of the radius r to the side x of the inscribed regular decagon is the famous golden section number

$$t = \frac{1 + \sqrt{5}}{2}$$

since clearly $t^2 = 1 + t$. We have thus shown that a regular pentagon and a regular decagon can be constructed with ruler and compasses. A good representation of a regular pentagon is simply obtained by taking a strip of paper, tying it into the simple trefoil knot, as shown in Fig. 1.4d, then carefully pulling it tight and flattening it.

Next we use Ptolemy's theorem to obtain the diagonals and the circumradius of a cyclic quadrilateral in terms of the sides a, b, c, d. Note first the sine law for triangles: If a, b, c, α, β, γ are the sides and the angles of a triangle and R is its circumradius, then

$$\frac{a}{\sin \alpha} = \frac{b}{\sin \beta} = \frac{c}{\sin \gamma} = 2R. \tag{27}$$

The proof follows immediately from Thales' theorem, as is shown in Fig. 1.5a, since the angles at A and at A' are both α. Further, the area S of the triangle is given by

$$S = \tfrac{1}{2}bc \sin \alpha;$$

hence by (27) we have

$$S = \frac{abc}{4R}. \tag{28}$$

We return now to our cyclic quadrilateral of sides a, b, c, d. If only these lengths are given but not their order, then there are three possible cyclic quadrilaterals since the side a can be opposite any one of the other three. All three quadrilaterals are shown in Fig. 1.5b; the original one,

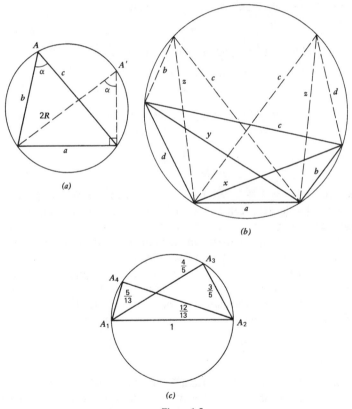

(a)

(b)

(c)

Figure 1.5

which has sides a, b, c, d in this order, and diagonals x, y, is drawn using a solid line. The other two cyclic quadrilateral have diagonals x, z and y, z, as shown in Fig. 1.5b. We apply Ptolemy's theorem to each one of the three cyclic quadrilaterals, getting

$$xy = ac + bd, \qquad xz = ad + bc, \qquad yz = ab + cd. \qquad (29)$$

Multiplying these three equations and taking square roots gives us xyz, and now x, y z are found to be

$$x = \sqrt{\frac{ac + bd}{ab + cd}}\,(ad + bc),$$

$$y = \sqrt{\frac{ac + bd}{ad + bc}}\,(ab + cd), \qquad (30)$$

$$z = \sqrt{\frac{ad + bc}{ac + bd}}\,(ab + cd).$$

Let R be the circumradius and S the area of any one of the three cyclic quadrilaterals of Fig. 1.5b. For instance, let us choose the one drawn in solid lines, and let us decompose it into two triangles, one with the sides a, b, x and the other with the sides c, d, x. Each triangle has R as its circumradius; applying (28) we find their areas to be

$$\frac{abx}{4R} \qquad \text{and} \qquad \frac{cdx}{4R}.$$

These two add up to the area S of the quadrilateral; hence

$$S = \frac{x}{4R}\,(ab + cd)$$

and so by (29)

$$S = \frac{xyz}{4R}. \qquad (31)$$

Finally, using in (31) the formula (11) for S and (30) for x, y, z we get a symmetric expression for R:

$$R = \frac{1}{4}\sqrt{\frac{(ab + cd)(ac + bd)(ad + bc)}{(s - a)(s - b)(s - c)(s - d)}}, \qquad s = \frac{a + b + c + d}{2}. \qquad (32)$$

In conclusion, we shall show as a consequence of Ptolemy's theorem that for any integer $n \geq 3$, n points can be found in the plane so that every two of them are an integer distance apart. This requires some comment. It is quite trivial to find such n points if they are allowed to lie on a straight line, but our n points will lie on a circle. Thus our proposition can be expressed as follows: for any $n \geq 3$ there is a cyclic n-gon with all sides and all diagonals of integer length.

To prove this, Ptolemy's theorem will be combined with the use of Pythagorean number triples u, v, w. These are integers u, v, w such that

$$u^2 + v^2 = w^2. \tag{33}$$

We verify that if p and q are any integers and

$$u = p^2 - q^2, \qquad v = 2pq, \qquad w = p^2 + q^2, \tag{34}$$

then (33) holds. The converse also holds: (33) implies (34) provided that the greatest common divisor of u, v, w is 1. Furthermore, we can take p and q in (34) to be of opposite parity. For instance, with $p = 2$ and $q = 1$ we get the very famous Pythagorean triple 3, 4, 5; with $p = 3$ and $q = 2$ we have 5, 12, 13. It is obvious that there are infinitely many distinct Pythagorean triples.

We take now a circle of diameter 1 and we let $A_1 A_2$ be a diameter, as in Fig. 1.5c. Next we place on our circle the point A_3 corresponding to the Pythagorean triple 3, 4, 5 and the point A_4 corresponding to the triple 5, 12, 13. In the cyclic quadrilateral $A_1 A_2 A_3 A_4$ all the distances $A_i A_j$ are rational except, possibly, the distance $A_3 A_4$. However, Ptolemy's theorem shows that $A_3 A_4$ also is rational. Next we place on our circle a fifth point A_5 again corresponding to a Pythagorean triple, and we use Ptolemy's theorem to show that all distances $A_i A_j$ are rational. Observe that each new Pythagorean triple actually gives us four new points. In this way we continue to add new points on the circle, keeping all distances rational. Once the number of points is n, we scale up our circle by an integer that is the least common multiple of all our distance denominators, and we are finished.

EXERCISES

1. Let P be a nonconvex n-gon. Show that by successive reflections of reentrant parts of P in lines through pairs of vertices we arrive at a convex n-gon. Can you estimate the maximal number of reflections needed, in terms of n?

2. Using Heron's formula alone, find the condition on three positive numbers a, b, c to be the sides of a triangle.

3. A quadrilateral Q has four of its sides fixed in length. Interpreting these as rigid rods and the vertices as hinges, we find that Q is not rigid, but adding one diagonal brace makes it rigid. Show that, similarly, any two diagonals rigidify a pentagon, and any three rigidify a hexagon. Conjecture and attempt to prove a generalization. Why is there a jump in difficulty from five to six vertices?

4. Prove that in a triangle

$$\cos \frac{A}{2} = \sqrt{\frac{s(s - a)}{bc}} \quad \text{and} \quad \sin \frac{A}{2} = \sqrt{\frac{(s - b)(s - c)}{bc}}.$$

5. In a triangle of sides a, b, c the length of the median connecting the midpoint of the side c to the opposite vertex is m. Show, using the cosine law, that $2a^2 + 2b^2 = 4m^2 + c^2$.

6. Show, using the previous problem, that in a parallelogram the sum of squares of the diagonals equals the sum of squares of the sides.

7. By using problem 5 several times show that in a convex quadrilateral the sum of squares of the sides equals the sum of squares of the diagonals plus four times the square of the distance between the midpoints of the diagonals.

8. Using problems 6 and 7 show that a convex quadrilateral is a parallelogram if and only if the sum of squares of the sides equals the sum of squares of the diagonals.

9. Show that the vector form of problem 6 is the vector identity

$$(\bar{a} + \bar{b})^2 + (\bar{a} - \bar{b})^2 = 2\bar{a}^2 + 2\bar{b}^2.$$

10. Show that with suitable notation the vector form of problem 7 is the vector identity

$$\bar{a}^2 + \bar{b}^2 + \bar{c}^2 + (\bar{a} + \bar{b} + \bar{c})^2 = (\bar{a} + \bar{b})^2 + (\bar{b} + \bar{c})^2 + (\bar{a} + \bar{c})^2.$$

(*Hint*: Let vectors $\bar{a}$, $\bar{b}$, $\bar{c}$ represent three consecutive sides; the fourth side is then $-\bar{a} - \bar{b} - \bar{c}$.)

11. Let the medians of a triangle have lengths x, y, z. Show that the area is

$$S = \tfrac{4}{3} \sqrt{s(s - x)(s - y)(s - z)}, \qquad s = \frac{x + y + z}{2}.$$

(*Hint*: Use problem 5.)

12. Find the area of a triangle in terms of its three heights. (*Hint*: Start with Heron's formula.)

13. Let T be an isosceles triangle with the vertex angle 20°. By reflecting T in its long sides so as to close up (to make a regular 18-gon) show by means of the isoperimetric theorem for the circle that the shortest curve that cuts T into two parts of equal area is not a straight segment. What is it?

14. Produce a quadrilateral, pentagon, hexagon, . . . in which all sides and all diagonals are of integer length. What can you conjecture about the minimal size of such an integral n-gon as n increases?

15. What are the extremes of the area of a convex quadrilateral of sides 2, 3, 4, 5?

16. A pentagon P has sides a, b, c, d, e in order. Its angles are adjusted so that P has maximum possible area, and x, y are then the diagonals from the vertex where a and b meet. Find the other three diagonals of P.

17. Prove the following generalization of Ptolemy's theorem:

$$x^2y^2 = a^2c^2 + b^2d^2 - 2abcd \cos(B + D)$$

(*Hint*: Use the diagram of Fig. 1.4*b* and apply the cosine law to $\triangle BDE$.)

18. Show that point E of Fig. 1.4*b* can always be found. Is it always so with respect to side d?

19. Show that a convex quadrilateral with sides a, b, c, d in order has an inscribed circle (i.e., one touching each side) if and only if $a + c = b + d$.

20. Show that if a convex quadrilateral of sides a, b, c, d has both an inscribed and a circumscribed circle, then its area is $\sqrt{abcd}$.

21. Let H be the regular heptagon of side 1 and let b be its shorter and c its longer diagonal. Use Ptolemy's theorem to show that

$$b^3 - b^2 - 2b + 1 = 0, \qquad c^3 - 2c^2 - c + 1 = 0$$

What does this suggest about constructing H?

22. By examining equations (24) and (26) what could you conjecture for a regular n-gon with n odd?

CHAPTER TWO
TRIANGLE
TRANSVERSALS

In this chapter we consider the relations between a triangle ABC and one or more of its transversals. Such a transversal is just a straight line crossing the sides of the triangle; however, a side is taken here in the extended sense: the whole straight line through two vertices of the triangle. Let us start with vertex transversals, that is, straight lines through a vertex of ABC. The first question is: How are such vertex transversals given? Consider a straight line AX, as in Fig. 2.1a, where X lies on BC or on the extension of BC beyond B or C. A natural way to fix the transversal is by means of the ratio BX/XC in which it divides the opposite side BC. Here we observe the convention that BX and XC are signed lengths: The sense from B to C, agreeing with the order $ABCA$, is positive and a segment on BC is positive or negative, depending on whether its sense agrees with that of BC or is opposite.

For instance, for the point X of Fig. 2.1a BX and XC are positive and so is the ratio BX/XC. For the point X_1, BX_1 is negative and X_1C is positive, so that the ratio BX_1/X_1C is negative and lies between -1 and 0. For the point X_2, BX_2 is positive and X_2C is negative and so the ratio BX_2/X_2C is negative and lies between $-\infty$ and -1.

In Fig. 2.1b the ratio BX/XC is plotted as X traverses the whole line BC. When X is on the interval BC, the ratio is positive and we then speak of internal division; when X is outside BC the ratio is negative and we then have external division. The ratio is 0 when $X = B$, it is 1 when X is the midpoint M of BC, and it is undefined as to sign but has infinite value when $X = C$. Limit considerations show that BX/XC approaches -1 when X moves indefinitely far to either side. This suggests the introduction of a convenient fictional "point at infinity" of the straight line BC, characterized by the ratio -1. Observe that there is just one such point at infinity on the line, not two, and that the vertex transversal through A, corresponding to that point at infinity, is then simply parallel to BC.

21

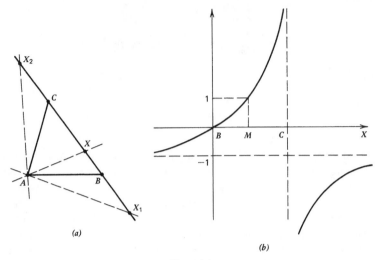

Figure 2.1

Suppose now that with the triangle ABC we have three vertex transversals AX, BY, CZ as shown in Fig. 2.2a, given by their ratios

$$BX:XC = l, \qquad CY:YA = m, \qquad AZ:ZB = n. \qquad (1)$$

Then several new triangles appear in the figure, and we are interested in their areas, relative to the area of the base triangle ABC.

The following expressions were given, without proofs, in 1891 by E. J. Routh, who needed them in the analysis of stresses and tensions in mechanical frameworks:

$$\frac{\text{Ar } XYZ}{\text{Ar } ABC} = \frac{lmn + 1}{(l + 1)(m + 1)(n + 1)}, \qquad (2)$$

$$\frac{\text{Ar } LMN}{\text{Ar } ABC} = \frac{(lmn - 1)^2}{(lm + l + 1)(mn + m + 1)(nl + n + 1)}. \qquad (3)$$

To prove these for the *general* situation of Fig. 2.2a is clearly a matter of some difficulty. We therefore simplify our work by employing the method of *special* cases. This begins with the time-honored mathematical phrase "without loss of generality it may be assumed that . . ." and the proof of the difficult general case is then reduced to proving a simpler special case only.

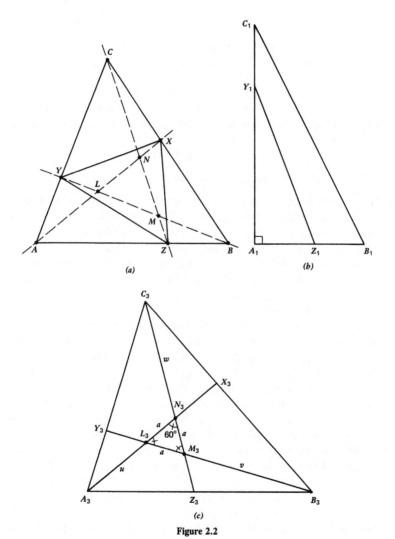

(a)

(b)

(c)

Figure 2.2

Here, as will be seen, we have the following: Without loss of generality it may be assumed that any one triangle in Fig. 2.2a is right-angled at any prescribed vertex, or that this triangle is isosceles, or even equilateral. Two questions arise:

1. Why may it be so assumed?
2. What good does it do us to assume it?

In answer to question 1 it is observed that only ratios of lengths and of areas enter in (2) and (3). Now such ratios do not change when the whole Fig. 2.2a is projected orthogonally onto another plane. If this operation projects points $A, B, \ldots, Z$ onto points $A_1, B_1, \ldots, Z_1$, then for any signed length UV and for any area Ar UVW we have

$$U_1V_1 = UV \cos \phi, \qquad \text{Ar } U_1V_1W_1 = \text{Ar } UVW \cos \theta.$$

Here θ is the angle of projection, that is, the angle between the two planes, and it is always assumed that $0° \leq \theta < 90°$. So, although lengths and areas themselves change under projection, their *ratios* stay the same. Moreover, the projection $A_1, B_1, \ldots, Z_1$ may be projected orthogonally again onto another plane as $A_2, B_2, \ldots, Z_2$, this may be further projected onto a third plane as $A_3, B_3, \ldots, Z_3$, and so on, and any ratios remain fixed.

After these remarks we can settle the first of our preceding questions. Let F stand for the whole configuration of Fig. 2.2a, and let UVW be any triangle in it. Let P be a horizontal plane, and start with F placed horizontally too, say, above P. In particular, the triangle UVW is initially horizontal. Consider its angle at U; if it is 90°, we do nothing. If it exceeds 90°, we rotate F about the bisector of the angle U. If it is less than 90°, we rotate F about the line through U perpendicular to that bisector. Since the projection of the angle U decreases in the first case down to 0° and increases in the second case up to 180°, we conclude by continuity that for a suitable projection the angle of $\Delta U_1V_1W_1$ at U_1 is 90°.

Next, we start again with $\Delta U_1V_1W_1$ horizontal, and we rotate about the shorter one of the sides adjacent to the right angle. For a suitable angle of tilt the second projection $\Delta U_2V_2W_2$ is both right-angled at U_2 and isosceles. Finally, we project for the third time, starting again with $\Delta U_2V_2W_2$ horizontal, and rotating about the angle bisector of the right angle at U_2. For a suitable angle of rotation the projected angle becomes 60°. However, the projected triangle is clearly always isosceles. Hence the third projection $\Delta U_3 V_3 W_3$, being an isosceles triangle with a 60° angle, is an equilateral triangle.

In answer to question 2 we remark that after the reduction to a right-angled or to an equilateral triangle, the ratios in (2) and (3) are very much easier to handle. In brief, the answer to question 2 is that it works.

Next, we prove (2). By a suitable orthogonal projection of ΔABC we obtain $\Delta A_1 B_1 C_1$, which is right-angled at A_1, as is shown in Fig. 2.2b. Then by (1)

$$\frac{AZ_1}{Z_1 B_1} = n, \qquad \frac{C_1 Y_1}{Y_1 A_1} = m$$

so that

$$\frac{A_1 Z_1}{A_1 B_1} = \frac{A_1 Z_1}{A_1 Z_1 + Z_1 B_1} = \frac{n}{n+1}, \qquad \frac{Y_1 A_1}{C_1 A_1} = \frac{Y_1 A_1}{C_1 Y_1 + Y_1 A_1} = \frac{1}{m+1}.$$

Therefore

$$\frac{\text{Ar } A_1 Z_1 Y_1}{\text{Ar } A_1 B_1 C_1} = \frac{A_1 Z_1}{A_1 B_1} \frac{Y_1 A_1}{C_1 A_1} = \frac{n}{(n+1)(m+1)}.$$

But the ratios of areas are invariant under projection, and so for Fig. 2.2a.

$$\frac{\text{Ar } AZY}{\text{Ar } ABC} = \frac{n}{(n+1)(m+1)}. \tag{4}$$

In the same way we prove the analogous equations

$$\frac{\text{Ar } CYW}{\text{Ar } ABC} = \frac{m}{(m+1)(l+1)}, \tag{5}$$

$$\frac{\text{Ar } BZX}{\text{Ar } ABC} = \frac{l}{(l+1)(n+1)}. \tag{6}$$

However, by a straightforward decomposition

$$\text{Ar } XYZ = \text{Ar } ABC - \text{Ar } AZY - \text{Ar } CYX - \text{Ar } BZY \tag{7}$$

and so by (4)–(7)

$$\frac{\text{Ar } XYZ}{\text{Ar } ABC} = 1 - \frac{n}{(n+1)(m+1)} - \frac{m}{(m+1)(l+1)} - \frac{l}{(l+1)(n+1)},$$

which is easily verified to give us (2).

Equation (3) is somewhat harder to prove. Our object is to find the ratio of areas of the transversal triangle *LMN* of Fig. 2.2*a* to the given triangle *ABC*. However, we succeed best by reversing the data: Start with *LMN* as given and work out to *ABC* from *LMN*. As was proved before, the triangle *LMN* may be assumed to be equilateral. Suppose that its sides have length *a*; we have now the configuration of Fig. 2.2*c* in which

$$A_3L_3 = u, \qquad B_3M_3 = v, \qquad C_3N_3 = w.$$

Next the transversal ratios *l*, *m*, *n* are expressed in terms of *u*, *v*, *w*, *a*. On account of the special values of our angles, which are 60° and 120°, the segments A_3L_3 and L_3B_3 are inclined at the same angle, namely, 60°, to C_3Z_3. Therefore, by setting up a simple proportionality we get

$$n = \frac{A_3Z_3}{Z_3B_3} = \frac{a + u}{v}$$

or

$$a + u = nv.$$

In the same way we get two further equations to form a system of three linear equations for the unknowns *u*, *v*, *w*:

$$u + a = nv, \qquad v + a = lw, \qquad w + a = mu. \tag{8}$$

Solving this for *u*, *v*, *w* we obtain

$$u = a\,\frac{nl + n + 1}{lmn - 1},$$

$$v = a\,\frac{lm + l + 1}{lmn - 1}, \tag{9}$$

$$w = a\,\frac{mn + m + 1}{lmn - 1},$$

recognizing already the expressions that appear in (3). Next on account of the 60° angles, we easily compute the areas of triangles

$$\text{Ar } L_3M_3N_3 = \frac{a^2\sqrt{3}}{4}, \qquad \text{Ar } A_3L_3B_3 = \frac{(a + v)u\sqrt{3}}{4},$$

$$\text{Ar } B_3M_3C_3 = \frac{(a + w)v\sqrt{3}}{4}, \qquad \text{Ar } C_3N_3A_3 = \frac{(a + u)w\sqrt{3}}{4}. \tag{10}$$

By a simple decomposition of $\Delta A_3B_3C_3$ we have

$$\text{Ar } A_3B_3C_3 = \text{Ar } L_3M_3N_3 + \text{Ar } A_3L_3B_3 + \text{Ar } B_3M_3C_3 + \text{Ar } C_3N_3A_3,$$

so that by (10)

$$\text{Ar } A_3B_3C_3 = \frac{[a^2 + (a + v)u + (a + w)v + (a + u)w]\sqrt{3}}{4}.$$

Together with the first equation in (10) we therefore have

$$\frac{\text{Ar } L_3M_3N_3}{\text{Ar } A_3B_3C_3} = \frac{a^2}{a^2 + (a + v)u + (a + w)v + (a + u)w}. \tag{11}$$

It is verified that the denominator may be rewritten in a symmetric form as

$$a^2 + (a + v)u + (a + w)v + (a + u)w$$
$$= \frac{1}{a}[(a + u)(a + v)(a + w) - uvw]$$

so that (11) becomes

$$\frac{\text{Ar } L_3M_3N_3}{\text{Ar } A_3B_3C_3} = \frac{a^3}{(a + u)(a + v)(a + w) - uvw}.$$

By means of (8) this can be expressed as

$$\frac{\text{Ar } L_3M_3N_3}{\text{Ar } A_3B_3C_3} = \frac{a^3}{(lmn - 1)uvw}.$$

Finally, substituting for u, v, w into the preceding equation from (9) and recalling that

$$\frac{\text{Ar } L_3M_3N_3}{\text{Ar } A_3B_3C_3} = \frac{\text{Ar } LMN}{\text{Ar } ABC},$$

we get (3).

Both in (2) and in (3) the three vertex transversals may be internal or external. In (2) it is supposed that l, m, and n are not -1. Here we recall that a vertex transversal has ratio -1 if and only if it is parallel to the opposite side of the triangle. In such case the denominator in the R.H.S. of (2) is 0 because the triangle XYZ is then undefined or, rather, because it has opened out into a semi-infinite strip (of infinite area). Similarly, the vanishing of the denominator in the R.H.S. of (3) means that some two transversals are parallel, and so the triangle LMN opens then out into a semi-infinite strip.

However, both in (2) and in (3) the ratios l, m, n may be 0 or infinite; the corresponding transversals coincide then with certain sides of the base triangle ABC. For instance, if l is infinite, then we get from (2) and (3)

$$\frac{\text{Ar } XYZ}{\text{Ar } ABC} = \lim_{l \to \infty} \frac{lmn + 1}{(l + 1)(m + 1)(n + 1)}$$

$$= \lim_{l \to \infty} \frac{mn + 1/l}{(1 + 1/l)(m + 1)(n + 1)} = \frac{mn}{(m + 1)(n + 1)}$$

and similarly

$$\frac{\text{Ar } LMN}{\text{Ar } ABC} = \lim_{l \to \infty} \frac{(lmn - 1)^2}{(lm + l + 1)(mn + m + 1)(nl + n + 1)}$$

$$= \frac{1}{(mn + m + 1)} \lim_{l \to \infty} \frac{lmn - 1}{lm + l + 1} \frac{lmn - 1}{nl + n + 1}$$

$$= \frac{1}{(mn + m + 1)} \lim_{l \to \infty} \frac{mn - 1/l}{m + 1 + 1/l} \frac{mn - 1/l}{n + (n + 1)/l}$$

$$= \frac{m^2 n}{(m + 1)(mn + m + 1)}.$$

Alternative proofs of (2) and (3) may be based on the elements of vector algebra. We sketch such a proof for (2). First, the transversal ratios are expressed in vector form: If vectors $\bar{U}$ and $\bar{V}$ correspond to points U and V in the plane, and if point W on UV divides it in the ratio $r:1$, then the vector $\bar{W}$ for the point W is

$$\bar{W} = \frac{\bar{U} + r\bar{V}}{1 + r}.$$

Conversely, every vector of this form corresponds to a point on UV. All this is simply verified, since

$$\bar{W} - \bar{U} = \frac{r}{1 + r}(\bar{V} - \bar{U}), \qquad \bar{V} - \bar{W} = \frac{1}{1 + r}(\bar{V} - \bar{U})$$

so that indeed

$$\bar{W} - \bar{U} = r(\bar{V} - \bar{W}).$$

In Fig. 2.2a let us introduce the vectors

$$\overline{AB} = \bar{u}, \qquad \overline{BC} = \bar{v}, \qquad \overline{AC} = \bar{u} + \bar{v}.$$

Then the vectors X, Y, Z corresponding to points X, Y, Z are

$$\bar{X} = \bar{u} + \frac{l}{l + 1}\bar{v}, \qquad \bar{Y} = \frac{1}{m + 1}(\bar{u} + \bar{v}), \qquad \bar{Z} = \frac{n}{n + 1}\bar{u}.$$

Therefore

$$\overline{ZX} = \bar{X} - \bar{Z} = \frac{1}{n+1}\bar{u} + \frac{l}{l+1}\bar{v},$$

$$\overline{ZY} = \bar{Y} - \bar{Z} = \left(\frac{1}{m+1} - \frac{n}{n+1}\right)\bar{u} + \frac{1}{m+1}\bar{v}.$$

Now, recalling the elementary properties of cross products, we have

$$\frac{\text{Ar } XYZ}{\text{Ar } ABC} = \frac{|\overline{ZX} \times \overline{ZY}|}{|\overline{AB} \times \overline{AC}|}$$

and since $\bar{u} \times \bar{u} = \bar{v} \times \bar{v} = 0$ we get

$$\frac{\text{Ar } XYZ}{\text{Ar } ABC} = \frac{1}{n+1}\frac{1}{m+1} - \frac{l}{l+1}\left(\frac{1}{m+1} - \frac{n}{n+1}\right)$$

$$= \frac{lmn + 1}{(l+1)(m+1)(n+1)}$$

giving us (2).

As was already mentioned, Routh obtained (2) and (3) in 1891 for the purpose of stress–strain analysis in mechanical frameworks. However, the chief feature of interest in (2) and (3) is probably that they provide far-reaching generalizations of two much earlier geometrical theorems. The first one goes back to antiquity and is due to the Greek geometer Menelaus of Alexandria, of the first century A.D. (no relation to the husband of Helen of Troy). The second theorem, although quite similar, was discovered much later, in 1678, by the Italian mathematician Giovanni Ceva (1647–1736)

Menelaus' Theorem. Three points X, Y, Z on the (extended) sides of a triangle ABC lie on a line if and only if $lmn = -1$.
Ceva's Theorem. Three vertex transversals AX, BY, CZ pass through a point if and only if $lmn = 1$.

The proofs follow immediately from (2) and (3) since the necessary and sufficient condition is that Ar $XYZ = 0$ in (2) and that Ar $LMN = 0$ in (3). We observe for Menelaus' theorem that if X, Y, Z lie on a line, so that $lmn = -1$, then either one or three of the ratios l, m, n are negative. Thus the Menelaus line XYZ either cuts all three sides externally or cuts one of them externally and the other two internally.

Suppose now that the three vertex transversals AX, BY, CZ pass

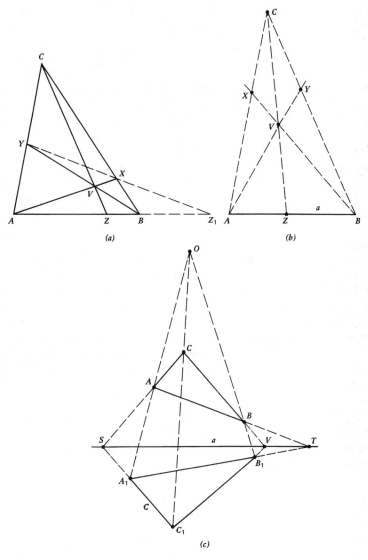

(a)

(b)

(c)

Figure 2.3

30

through one point V, as in Ceva's theorem. With reference to Fig. 2.3a let the line YX cut AB in Z_1. Then by Ceva's theorem we have

$$\frac{AZ}{ZB}\frac{BX}{XC}\frac{CY}{YA} = 1.$$

Since the points Y, X, Z_1 are on one line, we have by Menelaus' theorem

$$\frac{AZ_1}{Z_1B}\frac{BX}{XC}\frac{CY}{YA} = -1.$$

On comparing the last two equations it follows that

$$\frac{AZ}{ZB} = -\frac{AZ_1}{Z_1B}.$$

That is, the points Z and Z_1 divide the side AB in the same numerical ratio, one internally (with plus sign) and the other one externally (with minus sign). In particular, if Z is the midpoint of AB, we have $AZ/ZB = 1$ and for the point Z_1 we must then have $AZ_1/Z_1B = -1$. But this means that Z_1 is the point at infinity on AB; in other words YX is parallel to AB.

The following construction depends on the preceding: Given three points A, Z, B on a line a, such that Z is the midpoint of AB, and given a point X not on a, construct *by ruler alone* the line through X parallel to a. The construction is shown in Fig. 3.3b. Draw the line AX and choose on it any point C beyond X. Draw the lines CZ and XB, which intersect at point V. Then draw the line AV, which intersects CB at point Y. Now XY is parallel to a. More generally, given points A, Z, B on a line a, such that Z divides AB internally (or externally) in the ratio $r:1$, it is possible in the same way to find by ruler alone the point Z_1 on a that divides AB externally (or internally) in the ratio $-r:1$.

We shall next use the theorem of Menelaus to prove

Desargues's Theorem (Plane Case). If two triangles ABC, $A_1B_1C_1$ in the plane are such that the three lines AA_1, BB_1, CC_1 meet in a point, then the three pairs of corresponding sides, AB with A_1B_1, AC with A_1C_1, BC with B_1C_1, intersect in three points that lie on a line, and conversely.

This theorem may be expressed more transparently by using certain terms from the science of perspective. This science arose out of the attempts of Renaissance painters to study methods of representing the real three-dimensional world on the flat two-dimensional canvas.

Let F be a plane figure consisting of points A, B, C, ... and let F_1 be another figure in the plane of F, consisting of equally many corresponding points A_1, B_1, C_1, ... We say that F and F_1 are in point perspective if all the lines AA_1, BB_1, CC_1, ..., joining corresponding pairs of points, pass through a point O. This point is then called the center of perspective. The term *perspective* may be justified by noting that to the observer placed at O the two figures correspond point to point by the line of sight.

Such a visual approach is now continued: Our figures F and F_1 are said to be in line perspective if all pairs of lines such as AB with A_1B_1, AC with A_1C_1, and so on, intersect on a line a. This line a is then called the *axis of perspective*. Again we have a visual interpretation: An observer moving on the axis a can line up every pair of corresponding lines of the two figures.

The theorem of Desargues may be now briefly expressed:

Two triangles ABC, $A_1B_1C_1$ in the plane are in line perspective if and only if they are in point perspective.

As a matter of fact, the theorem of Desargues holds also for two triangles in space, not just in plane, but we shall prove only the plane case.

Let then ABC, $A_1B_1C_1$ be the two triangles, as shown in Fig. 2.3c, in point perspective, with the center of perspective at O. Let points S, U, T be the intersections of AC with A_1C_1, of BC with B_1C_1, and of AB with A_1B_1. We apply Menelaus' theorem three times:

Triangle	Menelaus line
ABO	A_1B_1T
BCO	B_1C_1U
CAO	C_1A_1S

This gives us the three Menelaus relations

$$\frac{AT}{TB}\frac{BB_1}{B_1O}\frac{OA_1}{A_1A} = -1,$$

$$\frac{BU}{UC}\frac{CC_1}{C_1O}\frac{OB_1}{B_1B} = -1,$$

$$\frac{CS}{SA}\frac{AA_1}{A_1O}\frac{OC_1}{C_1C} = -1.$$

To help us with writing these down, we may observe the three "skeletons"

$$\frac{A \quad B \quad O}{B \quad O \quad A} \quad \text{ for } \quad \Delta ABO,$$

$$\frac{B \quad C \quad O}{C \quad O \quad B} \quad \text{ for } \quad \Delta BCO,$$

$$\frac{C \quad A \quad O}{A \quad O \quad C} \quad \text{ for } \quad \Delta CAO.$$

Multiplying the three Menelaus relations and performing the cancellations, with some care as to signs, we obtain

$$\frac{AT}{TB} \frac{BU}{UC} \frac{CS}{SA} = -1.$$

But this means that TUS is a Menelaus line for ΔABC, and in particular the points T, U, S are on a line. Hence the triangles ABC and $A_1B_1C_1$ are in line perspective with the perspective axis TUS.

In the reverse direction, let us suppose that ΔABC and $\Delta A_1B_1C_1$ are in line perspective so that the points T, U, S lie on a line a. Let O be the intersection point of AA_1 with BB_1. The triangles ASA_1 and BUB_1 are now in point perspective, with the perspective center T. Therefore, as we have already proved, they are also in line perspective, and their perspective axis is OCC_1. This means that the points O, C, C_1 lie on a line. Hence the original triangles ABC and $A_1B_1C_1$ are in point perspective, with the center of perspective O. This completes the proof.

The criterion $lmn = 1$ of Ceva's theorem gives us a useful way of proving that many sets of three lines through the vertices of a triangle ABC must meet in a point. For instance, the three medians meet in a point, called the centroid of the triangle, because here $l = m = n = 1$. The three angle bisectors meet, in the circumcenter of the triangle, because then $l = c/b$, $m = a/c$, $n = b/a$. The three heights meet, in a point called the orthocenter of the triangle, because in this case the transversal ratios are $l = \cot B/\cot C$, $m = \cot C/\cot A$, $n = \cot A/\cot B$.

Somewhat more generally, let isosceles triangles with equal angles ω be built outward on the sides of the triangle ABC as shown in Fig. 2.4. We apply the sine law to the triangles ACC_1 and BCC_1, getting

$$\frac{CC_1}{\sin(A + \omega)} = \frac{AC_1}{\sin x}, \qquad \frac{CC_1}{\sin(B + \omega)} = \frac{BC_1}{\sin y}.$$

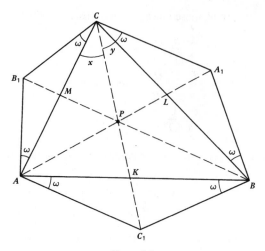

Figure 2.4

Since $AC_1 = BC_1$, because the triangle ABC_1 is isosceles, we get

$$\frac{\sin(A + \omega)}{\sin x} = \frac{\sin(B + \omega)}{\sin y}. \tag{12}$$

Apply similarly the sine law to the triangles AKC and BKC, getting

$$\frac{AK}{\sin x} = \frac{CK}{\sin A}, \qquad \frac{KB}{\sin y} = \frac{CK}{\sin B}.$$

Using these and (12) we conclude that

$$\frac{AK}{KB} = \frac{\sin(A + \omega) \sin B}{\sin(B + \omega) \sin A}.$$

The other two transversal ratios are similarly obtained:

$$\frac{BL}{LC} = \frac{\sin(B + \omega) \sin C}{\sin(C + \omega) \sin B}, \qquad \frac{CM}{MA} = \frac{\sin(C + \omega) \sin A}{\sin(A + \omega) \sin C}.$$

Hence it follows that

$$\frac{AK}{KB} \frac{BL}{LC} \frac{CM}{MA} = 1$$

and so by Ceva's theorem the straight lines AA_1, BB_1, CC_1 meet in a point P, as is shown in Fig. 2.4. This common point P may be relabeled as $P(\omega)$ to display its dependence on the angle ω.

To begin with, the point $P(\omega)$ is defined when $0 < \omega < 90°$. But a simple limiting procedure suggests that $P(0°)$ and $P(90°)$ are also defined: $P(0°)$ is the centroid of the triangle ABC, being the intersection of the three medians, and $P(90°)$ is the orthocenter, being the intersection of the three heights. It will be shown in a later section that $P(60°)$ is also an important point for the triangle ABC, its so-called Steiner point. It has the property of minimizing the sum of distances to the three vertices A, B, C. It may be recalled that the centroid $P(0°)$ has the property of minimizing the sum of the squares of those three distances.

We finish this section with some considerations that arise from its beginning. It will be recalled that we have mentioned the fictional "point at infinity" on a straight line. This gave us a more symmetric description of the transversal ratios. In particular, a segment on the line is divided externally by that line's point at infinity in the ratio -1. Further, we get a better view of the graph of Fig. 2.1b. The x-axis and the y-axis of that graph are the usual straight lines corresponding to our ordinary geometrical intuition.

Let us replace them by straight lines each of which is completed by its point at infinity. With such augmented lines as coordinate axes we have a complete 1:1 correspondence in Fig. 2.1b because now the point at infinity on the x-axis corresponds to the point labeled -1 on the y-axis. Conversely, the point at infinity on the y-axis corresponds to the point labeled C on the x-axis.

More generally, let us start with the Euclidean plane E, which is intuitively the model of ordinary elementary plane geometry. It contains points and straight lines with their usual properties. We now add to every straight line of E the point at infinity on that line. Are all such points at infinity distinct? No, because every family of parallel lines is completed by one and the same point at infinity. The totality of such points at infinity makes up one additional straight line, which we call the line at infinity. The plane E, augmented in this way, is called the projective plane P.

It may be verified that the properties of P are more symmetric that those of E. For instance, every two lines in P determine a unique point and every two points in P determine a unique line. This is not always true in E because here two parallel lines do not intersect and so have no common point. Both in E and in P infinitely many straight lines pass through every point. If the point is an ordinary one, that is, if it lies in E, then the lines through it have different directions. If the point is at infinity, then the lines through it are all parallel. In either case all such lines are intersected by the line at infinity, each line in one point.

It is possible to view P differently. Our previous remarks on the complete 1:1 correspondence of the graph in Fig. 2.1b suggest that the Euclidean straight lines are "open" whereas the projective straight lines are in some way "closed." In particular, a projective straight line is just a Euclidean line closed up by the addition of its point at infinity. To formalize these loose observations a little further, we can start with an ordinary Euclidean sphere S, for instance, with the locus given by the Cartesian equation $x^2 + y^2 + z^2 = 1$. Two opposite points on S, such as (x_1, y_1, z_1) and $(-x_1, -y_1, -z_1)$, are called antipodal. Equivalently, a straight line through the center of S cuts S in two antipodal points.

Let us now identify every two antipodal points of S by lumping them together. The result is a formalization of the projective plane P; its points are pairs of antipodal points of S, and its lines are great circles of S or rather, what is obtained from such great circles after the antipodal identification. There is now no distinction between ordinary and infinite elements, whether points or lines. An important difference between E and P is now easily noticed. The Euclidean plane is not compact, in the sense that we can produce infinite sequences of points in E without convergent subsequences. On the other hand, the projective plane P is compact, every infinite sequence of points in P having a convergent subsequence.

The preceding remarks are of course not intended as an introduction to projective geometry. Rather their aim is much more modest: to indicate how the addition to E of points at infinity and of the line at infinity gives us a greater symmetry of description, by avoiding special cases. Such cases arise in E precisely by the necessity of distinguishing between intersecting pairs and parallel pairs of lines.

An example is provided by the perspective formulation of Desargues' theorem. Here the center O of perspective of Fig. 3.3c can lie at infinity. We then have parallel perspective: The two triangles correspond point to point when viewed by the observer at infinity. Or the axis of perspective can be the line at infinity; the two triangles are then similar.

Another example will be given in the next section, on rotation and rolling. It will develop that every plane rigid motion is an instantaneous rotation about some center. This central point may vary from moment to moment. However, it can be objected that a translation along a straight line is certainly a rigid motion since it preserves all distances, yet it is not a rotation. But there is a way out of this problem because the center of rotation is now a point at infinity.

A third example will be met in the section on Pascal's theorem, concerning hexagons inscribed in conic sections. In any such hexagon the three pairs of opposite sides intersect in three points that are collinear. The pairs of opposite sides may be parallel, in which case some of their intersections are points at infinity. The line on which they lie may itself be the line at infinity. This example is historically (and otherwise) important, having started the subject of projective geometry.

EXERCISES

1. Construct an equilateral triangle with a vertex on each one of any three given parallel lines in the plane.

2. Does such an equilateral triangle exist when the three parallel lines are not in one plane?

3. Can one project an arbitrary triangle onto an equilateral triangle by *one* orthogonal projection?

4. Is the result of two consecutive orthogonal projections an orthogonal projection itself? (No, if it were so for two, the same would have been true for any number. A triangle could then be projected orthogonally onto another triangle with all three sides arbitrarily small, which is impossible. Why?)

5. In equations (2) and (3) let the three transversal ratios be equal: $l = m = n, = r$ say. Express Ar XYZ/Ar ABC and Ar LMN/Ar ABC in terms of r and simplify the fractions. If either of the area ratios is given, can r be found? What is r if Ar LMN/Ar $ABC = 1/7$?

6. If the three vertex transversals AX, BY, CZ pass through the point O, show that

$$\frac{OX}{AX} + \frac{OY}{BY} + \frac{OZ}{CZ} = 1.$$

7. Prove in detail that the angle bisectors of a triangle are concurrent.

8. Let the circle inscribed into the triangle ABC touch AB at Z, BC at X, and CA at Y. Show that the transversals AX, BY, CZ are concurrent.

9. Verify by drawing the existence of the points $P(30°)$, $P(45°)$, $P(60°)$. What angles appear to be subtended at $P(60°)$ by the sides of the triangle?

10. Prove equation (3) by vector methods (start by finding the vectors $\bar{L}$, $\bar{M}$, $\bar{N}$ corresponding to the points L, M, N in Fig. 2.2a: L lies both on AX and on BY, etc.)

11. Prove the following generalization of the theorem of Menelaus. Let $A_1A_2A_3 \ldots A_nA_1$ be an n-gon in the plane, $n \geq 3$, and let a transversal cut the side A_iA_{i+1} (or its extension) in X_i; we take $A_{n+1} = A_1$. Then

$$\frac{A_1X_1}{X_1A_2} \frac{A_2X_2}{X_2A_3} \cdots \frac{A_nX_n}{X_nA_1} = (-1)^n.$$

12. Conjecture and prove a similar generalization of Ceva's theorem (let n be odd, let C be any point in the plane of the n-gon, let the line A_iC divide the *opposite* side; the product of the n transversal ratios is 1).

CHAPTER THREE
ROTATION AND ROLLING

Let a circle C roll without slipping on a straight line and let P be any point on C. Then in the process of rolling, P describes a curve called the cycloid whose general appearance is shown in Fig. 3.1a. It is a periodic curve because the arch of Fig. 3.1a clearly repeats itself when the moving circle has rolled completely once over. The cycloid consists of such arches that are joined together, being vertically tangent at the cusp points. By the condition of rolling without slipping, the ratio of the base length of one arch to its height is the ratio of the circumference of C to its diameter, that is, π. The cycloid is a very famous curve, as is shown even by the multiplicity of its names: cycloid (from Greek *kuklos* = circle), roulette (from French, roll-path), trochoid (from Greek *trochos* = wheel). Famous mathematicians and scientists who have worked on the cycloid include R. Descartes (1596–1650), P. Fermat (1601–1665), G. de Roberval (1602–1675), B. Pascal (1623–1662), I. Newton (1642–1727), G. W. Leibniz (1646–1716), J. Bernoulli (1654–1705), C. Wren (1632–1723), and others. The last one was an architect whose interest in the cycloid was due to its role as arch in architecture.

The kinematic definition of *cycloid* enables us to visualize its shape empirically, for instance, as the path of a small light or a phosphorescent spot on the rim of a bicycle wheel, in the dark. Such an experiment brings out clearly the seeming breaks in the path due to the presence of the cusp points. Since *cycloid* comes to us from mechanics, we might expect the curve to have some special properties regarding motion. We mention two such properties, which explain partly the great early interest in this curve. Let an arch of the cycloid be turned upside down as in Fig. 3.1b; let A be the cusp point and B another point on the arch, at or before the lowest point (i.e., the midpoint of the arch). Suppose now that a heavy particle slides smoothly down the arc AB from A to B, starting from rest, under ordinary gravity. Then of all possible curves joining A to B the cycloid arc AB minimizes the time of sliding from A to B. This is the so-called brachistochrone property of the cycloid (from a Greek phrase meaning

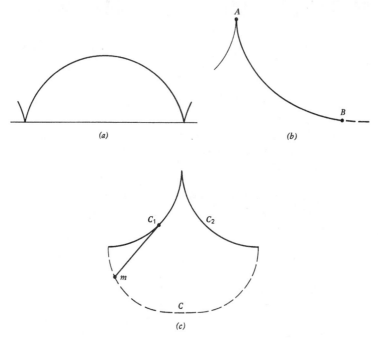

(a)

(b)

(c)

Figure 3.1

"shortest time"). It was the problem of finding the curve with the brach-istochrone property, solved first by Bernoulli and then by Newton, Leibniz, and others, that helped bring the cycloid into prominence. This problem also started the so-called calculus of variations, a branch of analysis devoted to the maxima and minima associated with the unknown curves, or surfaces, rather than with points on a given curve or surface.

Another famous property of the cycloid is connected with a problem of pendulum clocks. If a heavy mass m tied to a light string of length l is displaced by a small angle α off vertical and then allowed to oscillate, the oscillations have the period approximately equal $2\pi(l/g)^{1/2}$, where g is the acceleration of gravity. This approximation is better, the smaller the angle α. When α is large the approximation breaks down, giving the period values that are too large.

The problem arises now of producing a pendulum in which the period of oscillation is constant, no matter how large the displacement angle.

Such a pendulum is said to have the tautochrone property (from a Greek phrase meaning "equal time"). Suppose that the string of the pendulum is confined to move between two identical guard curves C_1 and C_2, as is shown in Fig. 3.1c. Then in its motion the string partly wraps itself along the guard C_1, then C_2, then C_1 again, and so on. As a result, the effective length of the pendulum is shortened, and the more so the bigger the displacement of mass m. Thus there is a compensation for the period of oscillation that grows with the displacement angle. It turns out that if C_1 and C_2 are upside-down cycloid arches and the total length of the string is half the length of one arch, then the motion of m is again along a cycloid arch C, as shown in Fig. 3.1c. Particularly important is the fact that this pendulum has the tautochrone property.

Along with the cycloid we have two other related curves: epicycloid and hypocycloid. These are defined in the same way as the cycloid except that now the circle C rolls without slipping on ($=$ epi) the outside, or the inside ($=$ hypo), of another circle. We derive now the equations of the three curves. For the cycloid we have the situation of Fig. 3.2a. The circle C of radius r rolls without slipping on the x-axis and the distinguished point of C is initially at the origin O. Some time later, C has rolled over to the position shown in Fig. 3.2a; it touches then the x-axis at B and its center is at A while the distinguished point is at P. The crucial mechanical condition of rolling without slipping is easily expressed: the arc PB of C has the same length as the straight segment OB. The convenient parameter to work with is the angle ϕ through which C has rolled: the rolling con-

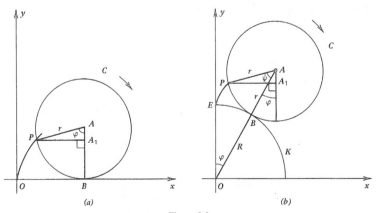

Figure 3.2

dition gives us $OB = r\phi$. Therefore the coordinates of A are $r\phi$ and r. Introduce the auxiliary right-angled triangle PAA_1; we get now coordinates x and y of P:

$$x = OB - PA_1, \qquad y = AB - AA_1.$$

Hence

$$x = r(\phi - \sin \phi), \qquad y = r(1 - \cos \phi) \tag{1}$$

are the parametric equations of the cycloid.

With the epicycloid we proceed in almost the same way, as is illustrated in Fig. 3.2b. Here K is the stationary circle of radius R, centered at the origin O. The rolling circle C of radius r was initially touching K at its highest point E. Some time later C has rolled over to the position illustrated in Fig. 3.2b; the distinguished point of C, initially at E, is now at P. Again we express first the condition that C rolls without slipping: arc EB of K and arc PB of C have the same length. In terms of the roll angles ϕ and ψ this gives us

$$r\psi = R\phi. \tag{2}$$

The coordinates of A are found to be

$$(R + r) \sin \phi, \qquad (R + r) \cos \phi.$$

Next we consider the same auxiliary right-angled triangle PAA_1 as before; its angle A is $\phi + \psi$. Therefore the coordinates x and y of P are obtained from the known coordinates of A:

$$x = (R + r) \sin \phi - r \sin(\phi + \psi),$$
$$y = (R + r) \cos \phi - r \cos(\phi + \psi). \tag{3}$$

Finally, we express ψ by ϕ from (2) and substitute into (3), getting

$$x = (R + r) \sin \phi - r \sin\left[\left(1 + \frac{R}{r}\right)\phi\right],$$
$$y = (R + r) \cos \phi - r \cos\left[\left(1 + \frac{R}{r}\right)\phi\right], \tag{4}$$

which are the parametric equations of the epicycloid.

To obtain the hypocycloid we could modify this procedure by placing the rolling circle C within the stationary circle K and otherwise keeping everything as before. However, it turns out that the equations of the hypocycloid are simply obtained from (4); we just replace r in (4) by $-r$. This has the effect of changing the rolling of C on the outside of K to rolling it on the inside, and we get

$$x = (R - r) \sin \phi - r \sin \left[\left(-1 + \frac{R}{r} \right) \phi \right]$$
$$y = (R - r) \cos \phi + r \cos \left[\left(-1 + \frac{R}{r} \right) \phi \right] \tag{5}$$

as the parametric equations of the hypocycloid.

The shape of this curve depends on the ratio R/r. Two special cases are of particular interest: $R = 2r$ and $R = 4r$. In the first case we get from (5)

$$x = r \sin \phi - r \sin \phi = 0, \qquad y = 2r \cos \phi.$$

Therefore this particular hypocycloid is just the straight segment $[-R, R]$ of the y-axis, that is, the vertical diameter of the circle K. This is a somewhat surprising way of generating straight-line motion by sole means of two circles. Observe here that the usual way of producing a straight line is by reference to some ruler or guide, i.e., to another straight line. Our two-circle method of generating straight segments was even of some practical importance. It was used in making optical diffraction gratings where many thousands of very closely spaced straight parallel scratches had to be drawn on a piece of glass.

When $R = 4r$ equations (5) give us

$$x = 3r \sin \phi - r \sin 3\phi, \qquad y = 3r \cos \phi + r \cos 3\phi. \tag{6}$$

To process these we recall De Moivre's formula

$$(\cos \phi + i \sin \phi)^n = \cos n\phi + i \sin n\phi, \qquad i^2 = -1,$$

which we use with $n = 3$. One thus obtains

$$\cos 3\phi = \cos^3 \phi - 3 \cos \phi \sin^2 \phi,$$

$$\sin 3\phi = 3 \cos^2 \phi \sin \phi - \sin^3 \phi.$$

With these we show that (6) becomes simply

$$x = R \sin^3 \phi, \qquad y = R \cos^3 \phi.$$

Since $\sin^2 \phi + \cos^2 \phi = 1$, we eliminate ϕ from the preceding pair of equations and get

$$x^{2/3} + y^{2/3} = R^{2/3} \tag{7}$$

as the equation of our special hypocycloid. The shape of this curve is shown in Fig. 3.3a. The curve is called astroid, from the Greek word meaning "star." Although the hypocycloids and epicycloids are less famous than the cycloid, they occur in a variety of situations, ranging from optics and astronomy to the design of geared teeth.

We refer now to the previously given list of early mathematicians who have worked on the cycloid. It can be seen from their dates that the

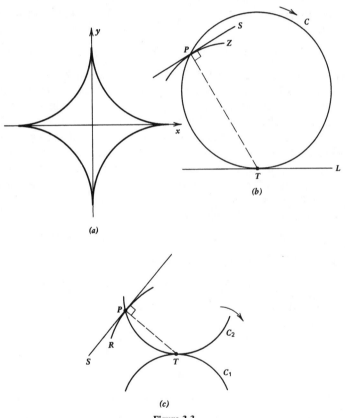

(a)

(b)

(c)

Figure 3.3

earlier work on the cycloid was being done before the development of
calculus. That is to say, those workers had no general machinery of limits
and derivatives. Nevertheless, they were able to cope with the basic
problem. This basic problem may be expressed geometrically as the problem of constructing the tangent to a given curve at an arbitrary point of
it. We shall see now how Descartes was able to handle tangents to the
cycloid seemingly without any use of the calculus machinery. The basic
idea is kinematic: The curve is regarded as the path of a moving point,
and the determination of the tangent to the curve at a point is converted
to finding the velocity of the moving point at an instant. Of course, the
cycloid from its very definition lends itself very well to such a kinematic
view. Starting with the cycloid and then generalizing, Descartes was led
to a principle that may be formulated as follows:

> The instantaneous motion of a plane rigid body moving in its own plane is
> a rotation about some point as center.

Why? Observe first that if the plane body C rotates (in its own plane),
then any two points of it remain the same distance apart. Thus the principle of Descartes is in a sense a converse of our observation: Because
C is rigid any two points of it stay at the same distance, and therefore the
instantaneous motion of C is a rotation about some center. However, this
center changes during the motion. Note that the phrase *instantaneous
motion* shows that we do not really avoid the use of limits. Rather, Descartes' principle enables us to transfer the use of limits from the analytical
machinery of derivatives into the geometric description of motion.

In one respect Descartes' principle might seem to fail: What happens
if the plane body C is being translated? Every point of it moves then with
the same velocity. How can the motion be regarded as a rotation? This
is simply taken care of if we recall the points at infinity mentioned in the
last section: A translation of C is a rotation about a point at infinity. This
mechanical interpretation shows even why any line L has only one point
at infinity and not two: The rotation about the hypothetical "right" infinite
point by a small angle α has the same effect as the rotation by $-\alpha$ about
the "left" infinite point.

Let us return now to the cycloid. As is shown in Fig. 3.3*b*, the circle
C rolls without slipping on the line L, and the point P, which is fixed on
the rim of C, is generating the cycloid Z. At a certain instant the situation
is as shown in the figure, the circle C then touches L at T and we ask,
How do we get the tangent line S to Z at the point P?

The condition that C rolls without slipping was already examined once but from a "static" point of view. The result was to translate this rolling condition into equality of lengths of certain two arcs. Now we express our rolling condition differently, kinematically. Since the line L on which C rolls does not move, the point T must be instantaneously at rest, for otherwise C is slipping. Therefore, by Descartes' principle, T is the instantaneous center of rotation. Therefore the instantaneous motion of P is at right angles to the segment PT. Hence the tangent S has been found: It is the straight line through P, perpendicular to PT. This can be generalized. Let a curve C_2 roll without slipping on a stationary curve C_1 as is shown in Fig. 3.3c. A fixed point of C_2 describes a curve R called the roulette of C_2 on C_1. We get thus a cycloid when C_2 is a circle and C_1 is a straight line, and an epicycloid or a hypocycloid when both C_1 and C_2 are circles. When C_2 touches C_1 at T, as in Fig. 3.3c, and the distinguished point of C_2 is at P, then the tangent S to the roulette R at P is perpendicular to PT.

We now sketch a proof of Descartes' principle. First, if the word *instantaneous* is omitted, it becomes briefly: Plane rigid motion is a rotation. Since the motion is rigid, it is enough to know the new positions, i.e., after the motion, of just two points A, B; let their new positions be A_1, B_1, as in Fig. 3.4a. Since the motion is rigid, $AB = A_1B_1$. Let I be the intersection of the perpendicular bisectors of the straight segments AA_1 and BB_1. If the segments are parallel, then I is the point at infinity and the motion is a translation. Otherwise, we show that the triangles AIB and A_1IB_1 are congruent (because $AI = A_1I$, $BI = B_1I$, $AB = A_1B_1$); therefore $\sphericalangle AIB = \sphericalangle A_1IB_1$ and so $\sphericalangle AIA_1 = \sphericalangle BIB_1 = \alpha$ say. It now follows that the new position A_1B_1 is the result of rotating AB through the angle α about I as center.

The case of instantaneous motion is taken care of just as we obtain derivatives from difference quotients; we let A_1 approach A. The limiting position of the center I is then the instantaneous center of rotation.

The usefulness of Descartes' principle will be shown on a standard and well-known calculus problem concerning maxima and minima: Let two straight corridors of widths a and b meet at right angles as shown in Fig. 3.4b. What is the length l of the longest straight segment that can be carried horizontally round the corner? We can imagine, for instance, a length of pipe, rail, or any stiff, thin rod being carried in the corridors. In attacking this problem we use a principle analogous to the one giving us the strength of a chain: It is determined by the weakest link. Although

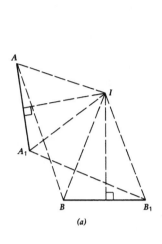

(a)

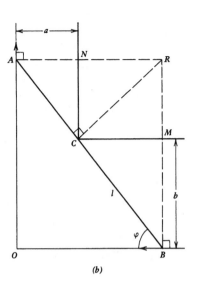

(b)

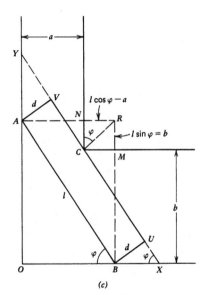

(c)

Figure 3.4

47

our formulation inquires after a maximum and uses the word *longest*, we reason that there is a critical or bottleneck position through which the rod must pass. What is really needed is the shortest segment AB of Fig. 3.4*b* passing through the corner C.

This will be done first by a straightforward application of calculus: Introduce the angle ϕ as shown, express the length y of the segment in terms of ϕ, and minimize. We get thus

$$y = a \sec \phi + b \operatorname{cosec} \phi, \qquad (8)$$

since $|AB| = |AC| + |CB|$. Differentiate y with respect to ϕ and set the derivative equal to 0, getting

$$\tan \phi = \left(\frac{b}{a}\right)^{1/3}. \qquad (9)$$

Now $\sec \phi$ and $\operatorname{cosec} \phi$ are expressed as

$$\sec \phi = \sqrt{1 + \tan^2 \phi}, \qquad \operatorname{cosec} \phi = \sqrt{1 + \frac{1}{\tan^2 \phi}}.$$

We use this and (9); substituting into (8) gives us the length l, which is the minimal value of y. After some simple computations we obtain a compact and symmetric equation

$$l^{2/3} = a^{2/3} + b^{2/3} \qquad (10)$$

which shows a close connection of our problem with the hypocycloid (7).

We now solve our problem again, by the use of Descartes' principle. Look first at the critical or bottleneck position of the segment AB in Fig. 3.4*b*. At this juncture the ends A and B are moving in the directions of the arrows. The Descartes principle tells us that the instantaneous motion of the segment AB is a rotation about the point R shown in Fig. 3.4*b*. How is the critical bottleneck position to be characterized now? We answer: CR is perpendicular to AB. For otherwise the segment AB will hit the corner C instead of sliding past it. If $\sphericalangle ACR > 90°$, then AB cannot move forward from the position of Fig. 3.4*b*; if $\sphericalangle ACR < 90°$, it cannot have arrived there. The rest is easy: Let l be the critical length and ϕ the critical angle. Then

$$l = a \sec \phi + b \operatorname{cosec} \phi \qquad (11)$$

while the condition $CR \perp AB$ gives us

$$\frac{CM}{MR} = \frac{l \cos \phi - a}{l \sin \phi - b} = \tan \phi \tag{12}$$

(for $\sphericalangle CRB = \phi$ so that $CM/MR = \tan \phi$). Now, substituting for l from (11) into (12) we get (9) again, and eventually the same expression (10) for l.

Even if it is granted that our new kinematic method of solving the problem gives us better geometrical insight, it is still possible to ask what else this new method can accomplish, as compared with the usual analytical method of calculus. Two generalizations of our problem will be considered.

Under the same conditions as before we ask: What is the greatest length l of a rectangle of width d that can be carried round the corner? Just as the previous problem corresponds to carrying horizontally a straight rod round the corner, so now we carry horizontally a rectangular table, say.

Again, we consider the critical, or bottleneck, position of the rectangle, as shown in Fig. 3.4c. As before, we find the instantaneous center R of rotation for the whole rectangle $ABUV$. We argue as before and find that $CR \perp UV$ is the characterization of the bottleneck position. By considering the rectangle $AOBR$ we find the lengths $l \cos \phi - a$ and $l \sin \phi - b$ shown in Fig. 3.4c, and hence we get (12) again. However, equation (11) no longer holds: the quantity $a \sec \phi + b \csc \phi$ is now XY, not l. Therefore, to obtain the desired maximum length l, we subtract XU and YV from XY and we get

$$l = a \sec \phi + b \csc \phi - d(\tan \phi + \cot \phi).$$

Hence the maximum length l and the critical angle ϕ are now given by the system of two equations

$$\frac{l \cos \phi - a}{l \sin \phi - b} = \tan \phi \tag{13}$$

$$l = a \sec \phi + b \csc \phi - d(\tan \phi + \cot \phi).$$

When the value for l is substituted from the second equation into the first one, l gets eliminated and we remain with the equation for ϕ:

$$\sin^2 \phi(a \sin \phi - d) = \cos^2 \phi(b \cos \phi - d). \tag{14}$$

The special case when $d = 0$ reduces to (9) just as it should. As it stands, (14) is a sextic equation for ϕ. A convenient way to show this is to use the standard substitution $\tan \phi/2 = t$, we get then a sixth-degree equation for t.

Our problem is thus solved but only in part; that is, we would have to apply now numerical techniques relying on an approximate solution of (14) for given numerical values of the parameters a, b, d.

Finally, we generalize our original problem by supposing that the two corridors of widths a and b meet at an arbitrary angle α. Clearly, it may be assumed that $0 < \alpha < 180°$. The kinematic method will be used to show that the maximal length l can be found by geometrical constructions with arbitrary accuracy. Geometrically, our problem is as shown in Fig. 3.5a: Given the angle $XOY = \alpha$ and a point C within the angle, to find the shortest straight segment through C whose endpoints lie on the arms of the angle. The point C is the corridor corner, and the widths of the corridors are a and b, as shown.

We must not expect to be able to construct our minimal segment by an elementary geometrical construction, that is, by using ruler and compasses only. Even for the right-angle case of $\alpha = 90°$, equation (9) shows that the critical angle ϕ satisfies

$$\tan \phi = \sqrt[3]{2}$$

when $b = 2a$. So if an elementary construction existed, we would be able to construct a segment of length $\sqrt[3]{2}$. This would constitute an elementary solution to the famous ancient Greek problem of duplicating a cube, which we discuss in a later section. But it is known from modern algebra that such a solution does not exist.

However, it is entirely reasonable to expect an approximate geometrical construction that would enable us to find the critical segment to within any prescribed accuracy. This is in complete analogy with numerical procedures, say, for evaluating integrals or solving differential equations. We may not be able to do it exactly in terms of known functions, but we have numerical algorithms that enable us to calculate what we want with arbitrarily good accuracy.

It is first necessary to obtain a geometrical characterization for our minimum problem. Let O_1 be the orthogonal projection of the vertex O on AB. Then

$$AB \quad \text{is minimum if and only if } AC = O_1B. \tag{15}$$

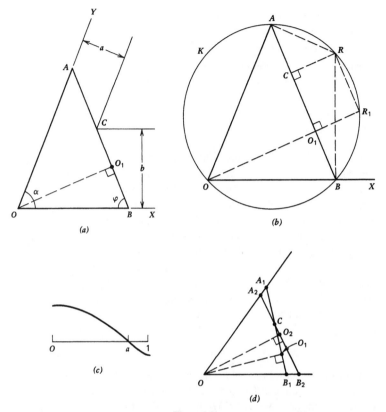

Figure 3.5

We use Fig. 3.5*b*. Suppose first that AB is minimum. Then, as we already know, R is the instantaneous center of rotation and $CR \perp AB$. If K is the circle circumscribing the quadrilateral $OARB$, then OR is a diameter, since $\angle OAR = 90°$. Let R_1 be the point on K such that $OR_1 \perp AB$ and let OR_1 cut AB at O_1. Since OR is a diameter of K, $\angle OR_1R = 90°$. Hence CO_1R_1R is a rectangle, since it is a quadrilateral three of whose angles are 90°. Therefore $AC = O_1B$.

Conversely, let AB be such that $AC = O_1B$. Let K be the circumcircle of $\triangle OAB$ and produce OO_1 to cut K at R_1. Let R be the point on K, such that $R_1R \parallel AB$. Then $\angle OR_1R = 90°$ and so OR is a diameter of K. Therefore $\angle OBR = \angle OAR = 90°$ so that R is the instantaneous center of rotation.

Finally, CO_1R_1R is a rectangle, since $AC = O_1B$, and therefore $CR \perp AB$. Hence AB is minimum, and so (15) is proved.

To describe our approximate geometrical construction more easily, we consider first the situation of Fig. 3.5c. Here we have the graph of a function $y = f(x)$ given for $0 \leq x \leq 1$, which is continuous, decreasing positive at $x = 0$, and negative at $x = 1$. It has therefore a unique root $x = a$ for which $f(a) = 0$. To find a we could proceed as follows. We test the sign of $f(1/2)$; if this is plus, then $1/2 < a \leq 1$, if this is minus then $0 \leq a < 1/2$ [we are not so foolish as to believe that $f(1/2) = 0$]. Thereafter we repeat the procedure: If $f(1/2)$ was >0 we test next the sign of $f(3/4)$ and if $f(1/2)$ was <0 we test the sign of $f(1/4)$, and so on. In n applications of such sequential testing the root a gets confined to an interval of length 2^{-n}. Thus a can be determined with arbitrarily good accuracy. This method of root localization is even of practical importance: It is relatively insensitive to errors, and it requires only very crude data (no functional values, no derivatives, but only signs).

Our approximate geometrical method is entirely analogous to the preceding. We use Fig. 3.5d and by trial and error we find two possibly close positions A_1B_1, A_2B_2 of segments passing through C. It is supposed that the following holds: When the vertex O is projected on A_1B_1 as O_1, and on A_2B_2 as O_2, then

$$A_2C < O_2B_2, \qquad A_1C > O_1B_1.$$

This of course corresponds to having $f(0)$ and $f(1)$ of opposite signs in our previous example. Our simple geometry guarantees the properties of continuity and monotonicity. We now "test at halfway" by taking A_3B_3, which passes through C and bisects the angle between A_1B_1 and A_2B_2. That is, we project O on A_3B_3, say as O_3, and we compare the lengths A_3C and O_3B_3. From here on the method is exactly as in our simple example of root localization.

EXERCISES

1. Without using their equations describe the general appearance of the epicycloid and the hypocycloid when $R = nr$ and n is a positive integer.

2. Repeat the same when $R = pr/q$ and p, q are positive integers whose greatest common divisor is 1. What happens when R/r is irrational?

3. The hypocycloid equations (5) were obtained in the text from the epicycloid equations (4) by changing r to $-r$. Obtain them directly following the pattern of deriving (4).

4. The problem of the longest rod that can be carried horizontally round the right-angle bend is changed so that instead of a right-angle bend we have (a) a T junction, (b) a cross. The widths of the corridors are still a and b. Is the greatest length increased?

5. Is the same true if the corridors meet at the angle α, $0 < \alpha < 90°$?

6. Attempt to solve the problem of maximum length when the corner C, of 90°, has been rounded off with radius r.

7. Two lines M and N meet at right angles and represent roads. A point X represents the entrance to a mine shaft; its distances from M and N are 300 and 500 ft. It is necessary to join M to N by the shortest straight road segment that is possibly close to x. However, the regulation states that such a road cannot be closer than 50 ft to X. Set up the problem geometrically and attempt to solve it.

8. Using the idea of the instantaneous center but without using the arc-length formula, show that the length of an arch of the cycloid is eight times the radius of the rolling circle. (*Hint*: Show that the required length L is given by

$$L = 2r \int_0^{2\pi} \sin \frac{\phi}{2}\, d\phi).$$

9. Prove the monotonicity property required in the last problem of this section.

CHAPTER FOUR

OBLIQUE SECTIONS
OF CERTAIN SOLIDS

Two solids are considered here, a cube and a torus, and in each case the interest is in finding a certain plane section of the solid by a particular oblique plane. The results obtained are of some interest in themselves. For instance, it is proved that a bona-fide hole of square cross section can be punched through a solid cube so that another cube, bigger than the original one, can be pushed through the hole. Also, it is shown that on the surface of a torus the number of circles that can be drawn through any fixed point, and that lie entirely on that surface, is not two but four. In reference to the title of this chapter, it may be recalled that the conics, that is, the important curves known as ellipses, parabolas, and hyperbolas, also happen to be oblique plane sections of a surface of a solid, the circular cone (as will be proved later).

However, our main purpose is to help develop the powers to see and appreciate geometrical relations in space. For three-dimensional though we are, we have been browbeaten into a flat-world view of things by the flood of paper, blackboards, billboards, and screens, and it is in many ways worth our while to become as stereoscopic as possible.

The first group of problems concerns the elementary geometry and the symmetry properties of an ordinary three-dimensional solid, the cube Q. Here by the symmetry of a solid we mean a quality of self-congruence or self-repetition. More precisely, symmetry is associated with certain motions called symmetry operations: Each such motion keeps the solid unchanged as a whole, through it may reshuffle parts of it among themselves.

For instance, if the axis a of the cube Q is taken as in Fig. 4.1a, then a rotation about a by 90°, or by an integer multiple of 90°, is a symmetry operation for Q. The mirroring in the plane, shown in Fig. 4.1b, is also a symmetry operation, and so is a different mirroring, in a diagonal plane, shown in Fig. 4.1c. A less obvious type of symmetry is illustrated in Fig.

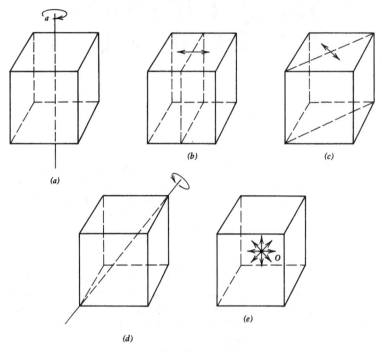

Figure 4.1

4.1*d*. Here the symmetry operation is a rotation by 120°, an angle not usually associated with the cube. This rotation is about the axis *t*, which contains the long diagonal of the cube, called also the body diagonal. By contrast, and for obvious reasons, the diagonals shown in Fig. 4.1*c* are called face diagonals. Finally, there is the type of symmetry shown in Fig. 4.1*e*: reflection in the center *O*. Here *O* is the center of the cube and the symmetry operation is as follows. *O* itself stays invariant, as it does in all other symmetry operations, and any other point *X* of the cube is shifted to X_1 where *O* is the midpoint of XX_1.

Let *R* denote the operation of Fig. 4.1*a*; *M* that of Fig. 4.1*b*, 4.1*c*, or 4.1*e*; and *T* that of Fig. 4.1*d*. Then four successive applications of *R* return the cube *Q* to its exact initial position point by point. The same is true for two applications of *M*, or three applications of *T*. All this is written down schematically in the algebraic terminology as

$$R^4 = M^2 = T^3 = 1.$$

The preceding remarks on symmetry will be finished by observing that it may be useful to distinguish the real, physically achievable motions, such as the rotations R or T, from the hypothetical "nonphysical" motions such as the mirrorings or the reflection in the center. The word *motion* may then be reserved for the physically achievable cases, and a new general term, *isometry*, is introduced. This comes from a Greek phrase meaning "same distance," and is defined as a distance-preserving point-to-point geometrical transformation. All symmetry operations become then special isometries. However, the difference between a motion and an isometry is not as big as it might appear. For instance, with the two-dimensional equivalent of Q, that is, a square S, a mirror reflection in a line or in the center is not achievable as a "real physical" motion in the plane of S, though it is achievable as a rotation in three dimensions. Similarly, the reflections of Fig. 4.1, though not achievable as three-dimensional motions, are in fact rotations of the cube in four dimensions.

The first problem concerning the cube is the following. Let Q be the cube of Fig. 4.2a and let P be the plane midway between A and H. That is, the plane P passes through the midpoint of the segment AH and is perpendicular to it; it is the locus of all points in space equidistant from A and H. The problem is the following: What is the figure $P \cap Q$ in which the plane P cuts the cube Q? Here the symbol $P \cap Q$ is borrowed from the standard terminology for sets: It is the collection of all points common to P and Q. For the purpose of visual training the plane P is deliberately left out of Fig. 4.2a. The idea is to follow the reasoning by reference to Fig. 4.2a as it is, and then to be able to imagine P, in relation to the figure, better and better, eventually to the point of seeing $P \cap Q$.

The figure $P \cap Q$ is some polygon, since it arises by the intersection of P and the plane faces of Q, and since P cuts every face of the cube in a straight segment. By its definition as the locus of points equidistant from A and H the plane P contains the midpoint of each one of the six edges of the cube drawn in Fig. 4.2a in heavy line. Hence $P \cap Q$ is a hexagon. What sort of a hexagon?

By using the congruence of certain obvious triangles it is easy to show that any two consecutive sides of the hexagon $P \cap Q$ are of the same length. Further, those two consecutive sides make the same angle as any other two consecutive sides. Thus $P \cap Q$ is a hexagon all of whose sides, and all of whose angles, are equal. It is therefore a regular hexagon.

It may be a help in visualizing this regular hexagon to consider how the cube Q of Fig. 4.2a is cut by a variable plane that moves across Q

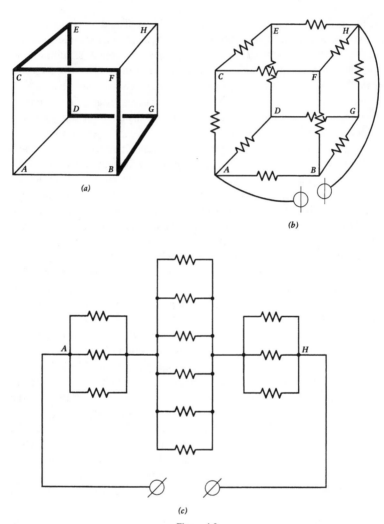

(a)

(b)

(c)

Figure 4.2

from A to H, remaining at all times perpendicular to AH. In other words, we are interested in how the regular-hexagon section *develops*. Denote the moving plane by $P(X)$: it passes through a point X on AH; the variable section is then $P(X) \cap Q$. When X is the point A, this section is just the point A itself. When X starts moving toward H the section $P(X) \cap Q$ is

Figure 4.3

an equilateral triangle that grows in size. As is shown in Fig. 4.3, at a certain instant the moving plane $P(X)$ hits all three vertices B, C, D of the cube Q at once, and the section $P(X) \cap Q$ is then the equilateral triangle BDC. When X moves on, the section changes in character: from an equilateral triangle it changes to a hexagon obtained by a symmetric corner truncation of three small equilateral triangles. The amount truncated grows as X moves toward the midpoint of AH. When X arrives at that midpoint the section becomes precisely our regular hexagon $P \cap Q$. As X moves past the middle of AH toward H, the same series of sections is obtained, but they repeat themselves in reverse, and with triangles pointing down rather than up. This reversal is associated with the symmetry of reflecting the cube in its center, shown in Fig. 4.1e. The triangular sections $P(X) \cap Q$ pointing up occur when X describes the first one-third of AH, the hexagonal sections correspond to the middle one-third, and the triangles pointing down to the last one-third.

There is the following convincing demonstration that the section $P \cap Q$ is a regular hexagon. Take the cube Q of Fig. 4.2a and make seven cuts: along the edges AB, AC, AD; CF; EH, FH, GH. Now the surface of the cube is made up of six squares connected by the remaining five edges, and it can be flattened out onto the plane. The result is a flat face net of the cube, shown in Fig. 4.4a. The boundary of the hexagon $P \cap Q$ appears here as a single straight segment UV joining the midpoint U of CF in the upper square $ABFC$ to the midpoint V of CF in the lower square $CFHE$. A model of the cube Q with the hexagon $P \cap Q$ on it can be easily made as follows. The net of Fig. 4.4a with the segment UV drawn on it is cut out, for instance, in thin cardboard. Then the six squares are bent along the five joining edges BF, BG, GD, DE, EC, and finally we join the square faces along the seven edges we have cut. However, the surface of the cube Q has also other face nets; for instance, the one shown in Fig. 4.4b; here the hexagon $P \cap Q$ appears as a broken line with three parts. Of course, when the whole cube is correctly glued up, those three parts join up to form the regular hexagon $P \cap Q$ again.

An interesting consequence can be deduced from the cube's face net

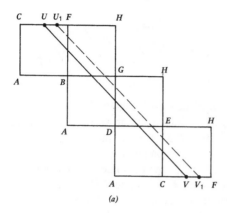

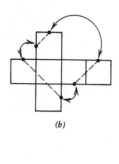

(a)

(b)

Figure 4.4

of Fig. 4.4a. Consider all the hexagonal sections $P(X) \cap Q$ of Fig. 4.3. Then all of them have the same circumference. For it is simple to show that in the planar development of Fig. 4.4a any such hexagon appears as some straight segment $U_1 V_1$ obtained by a mere parallel translation of UV.

We have proved that the hexagon $P \cap Q$ is regular by showing the equality of its six sides and of its six angles. The proof of this is particularly simple by working with the face net of Fig. 4.4a. The question arises whether $P \cap Q$ could be shown to be a regular hexagon by symmetry arguments alone. This question is justified since the symmetry properties may be considered more basic than the metric properties. We start by observing that both the cube Q and the plane P are kept unchanged by the symmetry operations T and M_2. With reference to Fig. 4.2a, T is the rotation by 120° about AH, and M_2 is the mirror-reflection in the plane $AGHC$. The symmetry operations T and M_2 are illustrated by Fig. 4.1d and Fig. 4.1c.

Since both P and Q are invariant under T and M_2 the same invariance applies to their common part, that is, the section $P \cap Q$. Does this invariance force $P \cap Q$ to be a regular hexagon? No, only to be semi-regular: the triples of alternating sides are of equal length, and all angles are 120°. We note in fact that every section of Fig. 4.3, whether hexagonal or triangular, is also unchanged by both T and M_2. However, the cube Q and the plane P are also invariant under the center reflection M_3 of Fig. 4.1e. Therefore so is $P \cap Q$; and *now* $P \cap Q$ must be a regular hexagon. That is to say, the symmetry operation M_3 suffices to distinguish the regular hexagon $P \cap Q$ among all sections of Fig. 4.3a.

The determination of $P \cap Q$ by means of a symmetry argument helps us to solve another and seemingly quite different problem. Let 12 identical 1-ohm resistances be connected together in the electrical network of Fig. 4.2b so as to form the edges of a cube. A potential difference is then applied across the terminals A and H, and the problem is, What is the total resistance of this cubical network?

The difficulty here is that our electrical network does not arise by simple series and parallel groupings of its 12 elements. We recall that if resistances R_1, R_2, ... R_n are connected together and R is the total resistance, then

$$R = R_1 + R_2 + \cdots + R_n$$

for the series connection, and

$$\frac{1}{R} = \frac{1}{R_1} + \frac{1}{R_2} + \cdots + \frac{1}{R_n}$$

for the parallel connection. Although our cubical network is not series-parallel, the symmetry operation T of rotating about AH is used to reduce it to a simple network. This rotation T by 120° permutes the vertices, or terminals, B, C, D among themselves, and also F, E, G among themselves. Therefore there is *electrical* as well as *geometrical* symmetry. In particular, the points B, C, D and the points F, E, G are at the same electrical potential, so that each triple can be lumped together without changing anything electrically. One gets thus the electrically equivalent but much simpler network of Fig. 4.2c. This is of the series-parallel type, and the total resistance is now easily found to be $\frac{5}{6}$ ohm.

The last one of our problems associated with the cube is, how do we put a square hole through the smaller one of two cubes so that the bigger one can be pushed through it? This might appear to be impossible but let us consider first the analogous situation in two dimensions. Given a unit square made of stiff, thin cardboard, let a straight diagonal slit be made in it, as is shown in Fig. 4.5c. It is clear now that another square of side $< \sqrt{2}$ can be pushed through for a suitable slit. But it may be objected, and validly too: that is not a correct analog because here the larger square is taken out of the plane of the smaller one, whereas we cannot do this for our two cubes. As will be seen, the correct plane analog is that of Fig. 4.5d, where the larger square, of side $< \sqrt{2}$, and the smaller one,

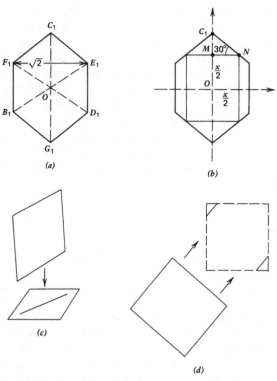

Figure 4.5

are in the same plane. This analog may even suggest how to attack the three-dimensional problem: by cutting a square hole through the smaller cube along the body diagonal.

Let our cube Q of Fig. 4.2a have unit edges and let us begin by projecting Q onto a plane perpendicular to AH; any such plane will do. The projection is, as may perhaps be guessed by now, a regular hexagon, and the symmetry argument using the rotation T will prove it. How big a regular hexagon is this projection? The face diagonals EG, GF, FE, all of length $\sqrt{2}$, are parallel to the plane onto which we project. Also, the points E, F, G project as alternate vertices E_1, F_1, G_1 of our hexagon, as is shown in Fig. 4.5a. Therefore the size of the regular hexagon is known: The distance between any two alternate vertices is $\sqrt{2}$, as is shown in Fig. 4.5a. Hence the distance of any vertex to the center O is

$\sqrt{2/3}$. We ask now, how big a square can be inscribed into our regular hexagon? Suppose that a square is taken as shown in Fig. 4.5b, let N be its vertex and x its side. Since $OC_1 = \sqrt{2/3}$ we have

$$C_1M = \sqrt{\frac{2}{3}} - \frac{x}{2}, \quad \tan 30° = \frac{C_1M}{MN}.$$

But $MN = x/2$ and $\tan 30° = 1/\sqrt{3}$. Therefore

$$\frac{\sqrt{2/3} - x/2}{x/2} = \frac{1}{\sqrt{3}}$$

and solving for x we find that

$$x = \sqrt{6} - \sqrt{2} = 1.0353 \ldots.$$

It follows now that by punching a suitably placed hole of square cross section through a cube we can push through it any cube of size up to 3.53 . . . % bigger than the first one.

Having finished with the cube, we go on now to the torus. This is defined to be a solid of revolution obtained by rotating a circle C about a line L that lies in the plane of C but does not cross it. Actually, only the torus *surface* T is of interest, and the following question is raised:

How many circles lying entirely on T can be drawn through an arbitrary given point p of T?

For ease of orientation let the torus T always be placed on a horizontal plane. Then two obvious circles pass through p:

1. The vertical (meridional) circle C_V.
2. The horizontal (latitude) circle C_H.

The names in the brackets are derived by analogy to the meridians and latitude circles of the terrestrial sphere. The vertical circle C_V is a position of the generating circle C, as it is revolved around the axis L, when it passes through p. The horizontal circle C_H is traced out by a point q of C; that point q is taken that will pass through p during the revolution. Yet, the surprising answer, which is sometimes missed even by old hands at the game, is that there are not two but *four* such circles on T through p.

Let the revolving circle C have radius r and let its center be at the distance a from the origin O in the vertical XZ-plane, as is shown in Fig.

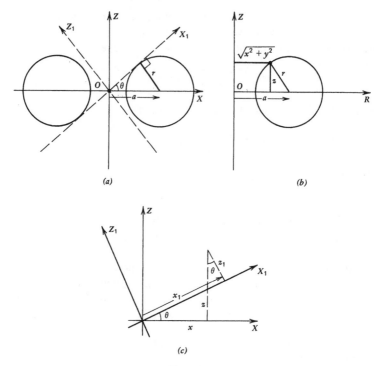

Figure 4.6

4.6a; it is always assumed that $a > r$. The axis of revolution, that is, the line L, is the Z-axis. The Y-axis passes through the origin O and is at right angles to the plane of the figure. Then the Cartesian equation of the torus T in the XYZ-coordinate system is

$$(a - \sqrt{x^2 + y^2})^2 + z^2 = r^2. \tag{1}$$

Here $\sqrt{x^2 + y^2}$ is the distance from the point (x, y, z) on the torus to the Z-axis, and (1) follows from the geometry of Fig. 4.6b in which R represents any line through O in the XY-plane.

Let new axes X_1 and Z_1 be introduced in the XZ-plane as shown in Fig. 4.6a. This amounts to rotating the XZ-coordinate system through the angle θ given by

$$\sin \theta = \frac{r}{a}. \tag{2}$$

Therefore the old coordinates (x, z) and the new ones are related by the rotation formulas

$$x = x_1 \cos \theta - z_1 \sin \theta, \qquad z = x_1 \sin \theta + z_1 \cos \theta, \qquad (3)$$

as is simply shown by taking projections in Fig. 4.6c. Since a point has the same distance to the origin in either system, it follows that

$$x^2 + z^2 = x_1{}^2 + z_1{}^2$$

or, in terms of (3),

$$(x_1 \cos \theta - z_1 \sin \theta)^2 + (x_1 \sin \theta + z_1 \cos \theta)^2 = x_1{}^2 + z_1{}^2. \qquad (4)$$

The axes X_1 and Z_1 are completed by adding the third axis Y_1, which merely coincides with the old Y-axis; this means that $y = y_1$. There are now two complete Cartesian coordinate systems XYZ and $X_1 Y_1 Z_1$. The equation of the torus T which is (1) in the XYZ-system, becomes in the new system

$$(\sqrt{(x_1 \cos \theta - z_1 \sin \theta)^2 + y_1{}^2} - a)^2 + (x_1 \sin \theta + z_1 \cos \theta)^2 = r^2.$$

By squaring out, simplifying, and squaring again to get rid of the square root, this is put first in the form

$$(x_1{}^2 + y_1{}^2 + z_1{}^2 + a^2 - r^2)^2 = 4a^2[y_1{}^2 + (x_1 \cos \theta - z_1 \sin \theta)^2]$$

and then, observing (4), in the form

$$(x_1{}^2 + y_1{}^2 + z_1{}^2 - a^2 - r^2)^2 = 4a^2[r^2 - (x_1 \sin \theta + z_1 \cos \theta)^2]. \qquad (5)$$

Now comes the crucial question: In what plane curve does the $X_1 Y_1$-plane cut the torus T? To get its equation we put $z_1 = 0$ in (5) and use (2) to express $\sin \theta$. After some further algebra one gets

$$(x_1{}^2 + y_1{}^2 + r^2 - a^2)^2 - 4r^2 y_1{}^2 = 0.$$

The L.H.S. is factorizable being the difference of two squares and the above becomes

$$[x_1{}^2 + (y_1 - r)^2 - a^2][x_1{}^2 + (y_1 + r)^2 - a^2] = 0.$$

Hence comes the unexpected result: the section of the torus T by the $X_1 Y_1$-plane consists of two circles C_1 and C_2 with the equations

$$C_1: x_1{}^2 + (y_1 - r) = a^2; \qquad C_2: x_1{}^2 + (y_1 + r)^2 = a^2.$$

These circles C_1 and C_2 are sometimes called, after their discoverer, the

Villarceaux circles. Unlike the previous pair C_V and C_H, the new circles lie obliquely on the horizontal torus T. From the definition of T as a surface of revolution it follows that C_1 and C_2 can slide over T sweeping out every point of T. Equivalently, either circle C_1 or C_2 generates all of T by being rotated once about the Z-axis.

It has therefore been established that through every point p of the torus there pass *at least* four circles: a vertical circle C_V, a horizontal circle C_H, and two oblique circles C_1 and C_2 (with a slight shift of notation).

It still remains to be shown that there are no others. Since the circle is a plane curve it follows that every circle on the torus T arises as an intersection of T with a certain plane P. On account of the rotational symmetry of T, every plane section of T has the equation of the form (5) with z_1 put equal to some constant c and with a suitable angle θ. We therefore have to show that the only circles represented by the equation

$$(x^2 + y^2 + c^2 - a^2 - r^2)^2 - 4a^2[r^2 - (x \sin \theta + c \cos \theta)^2] = 0 \quad (6)$$

are of the four types: C_V, C_H, C_1, or C_2. Suppose then that a circle K is represented by (6). This K has the equation of the form

$$(x - p)^2 + (y - q)^2 - r_1{}^2 = 0 \quad (7)$$

and therefore the L.H.S. of (7) divides the L.H.S. of (6). Since the one is a quadratic and the other a quartic, we must have

$$(x^2 + y^2 + c^2 - a^2 - r^2) - 4a^2[r^2 - (x \sin \theta + c \cos \theta)^2] \quad (8)$$

$$= [(x - p)^2 + (y - q)^2 - r_1{}^2][Ax^2 + Bxy + Cy^2 + Dx + Ey + F].$$

Comparing the coefficients of x^4 and y^4 on both sides shows that $A = C = 1$, also $B = 0$ since there are no x^3y or xy^3 terms in the L.H.S. But there are no cubic terms in L.H.S. either, and so (8) becomes

$$(x^2 + y^2 + c^2 - a^2 - r^2)^2 - 4a^2[r^2 - (x \sin \theta + c \cos \theta)^2]$$

$$= [(x - p)^2 + (y - q)^2 - r_1{}^2][(x + p)^2 + (y + q)^2 - r_2{}^2]. \quad (9)$$

This already gives some information: Any plane P that cuts the torus T in a circle, cuts it also in another circle, and the two circle centers are symmetric with respect to the origin. Next the absence of the y-term in the L.H.S. of (9) implies that

$$q(r_2{}^2 - r_1{}^2) = 0$$

and since the radii r_1 and r_2 are positive, there are two cases to consider: either $q = 0$ or $r_1 = r_2$. In the first case $q = 0$ and so (9) is

$$(x^2 + y^2 + c^2 - a^2 - r^2)^2 - 4a^2[r^2 - (x \sin \theta + c \cos \theta)^2]$$
$$= [(x - p)^2 + y^2 - r_1{}^2][(x + p)^2 + y^2 - r_2{}^2]. \qquad (10)$$

When the coefficients of x^2 and y^2 terms are compared it is found that

$$a^2 \sin^2 \theta = -p^2,$$

which implies that $p = 0$ and $\sin \theta = 0$. Therefore the plane P cuts the torus T in the locus

$$[x^2 + y^2 - r_1{}^2][x^2 + y^2 - r_2{}^2] = 0$$

that is, in two circles of the type C_H. The plane P is horizontal since $\theta = 0$.

Suppose next that $r_1 = r_2, = R$ say. Then (9) becomes

$$(x^2 + y^2 + c^2 - a^2 - r^2)^2 - 4a^2[r^2 - (x \sin \theta + c \cos \theta)^2]$$
$$= [x^2 + y^2 + p^2 + q^2 - R^2 - 2px - 2qy] \qquad (11)$$
$$\times [x^2 + y^2 + p^2 + q^2 - R^2 + 2px + 2qy].$$

The R.H.S., being a product of the sum and the difference, is the difference of squares so that (11) is of the form

$$(x^2 + y^2 + c^2 - a^2 - r^2)^2 - 4a^2[r^2 - (x \sin \theta + c \cos \theta)^2]$$
$$= (x^2 + y^2 + p^2 + q^2 - R^2)^2 - 4(px + qy)^2. \qquad (12)$$

Comparing the coefficients of x-terms and xy-terms, we find that

$$c \sin \theta \cos \theta = 0 \qquad \text{and} \qquad pq = 0.$$

The case $q = 0$ has already been shown to lead to C_H circles; suppose therefore that $p = 0$. If $\sin \theta = 0$, the plane P cutting T is horizontal and we get C_H circles again, as sections. If $\cos \theta = 0$, then $\theta = 90°$ and P is vertical, and it cuts T in the locus

$$[x^2 + (y - q)^2 - R^2][x^2 + (y + q)^2 - R^2] = 0$$

as is seen by factorizing the R.H.S. of (12) with $p = 0$. The sections are now the vertical circles C_V. Finally, there remains the possibility that c

$= 0$ and $p = 0$. Then (12) may be written as

$$(x^2 + y^2 + r^2 - a^2)^2 + 4x^2(a^2 \sin^2 - r^2) - 4r^2y^2$$
$$= (x^2 + y^2 + q^2 - R^2)^2 - 4q^2y^2.$$

Therefore the coefficient of x^2 vanishes, or

$$a \sin \theta = \pm r.$$

Either sign gives the same result: the two oblique Villarceaux circles C_1 and C_2.

   This completes the proof. We have proved that through every point of a (horizontally placed) torus T there pass exactly four circles: one horizontal circle C_H, one vertical circle C_V, and two oblique circles C_1 and C_2. Their exact relationship to each other may be explained by using some terms from the elementary topology of surfaces. In particular, we are interested in the deformability of various curves on T.

   First, there is the obvious fundamental difference between the surface S of a sphere and our torus T, arising from the fact that T has a hole through it. The surface S is simply connected: Any closed curve lying entirely on S can be continuously shrunk to a point, remaining on S at all times. On the other hand, T is not simply connected: Not every closed curve on T can be continuously shrunk on T to a point. For instance, neither C_H nor C_V (nor for that matter, C_1 or C_2) can be deformed on T to a point. Also, C_V and C_H cannot be deformed on T into each other.

   For that reason, C_V and C_H are called independent cycles of T. Using C_V and C_H as the basis, we can describe all possible types of curve winding for closed curves on T that pass through our fixed point p. For instance, $2C_V$ would describe the winding type of a curve that goes twice around T vertically intersecting itself in the process, $-3C_H$ refers similarly to curves that go about T three times horizontally and in the sense opposite to that we have chosen as positive for C_H, and so on. Since C_V and C_H have a positive and negative sense, any winding type of a closed curve on T through p is of the form

$$mC_V + nC_H, \qquad m, n = \ldots, -1, 0, 1, \ldots.$$

The type $m = 0$, $n = 0$ means that the curve in question winds around T neither vertically nor horizontally; it is therefore a *contractible* curve and can be shrunk on T to the point p. Now we find that our four circles on the torus are of the following winding types

$$C_V: 1, 0; \qquad C_H: 0, 1; \qquad C_1: 1, 1; \qquad C_2: 1, -1.$$

That is, contractible curves 0, 0 cannot be circles, $|m|$ and $|n|$ must be ≤ 1 because otherwise the curves are too twisted to be even planar, and there is no distinction between the type (m, n) and $(-m, -n)$. Four types remain then, and there is exactly one circle per type.

Together with the cube Q and its regular hexagon section $P \cap Q$, it was found useful to consider also the sections $P(X) \cap Q$ of the cube by planes parallel to P. These parallel sections, shown in Fig. 4.3, illustrate just how the regular-hexagonal section develops. The same will be now done for the torus T; a counterpart of Fig. 4.3 will be obtained showing us how the special bicircular section consisting of C_1 and C_2 develops. Let $P(c)$ be the plane $z_1 = c$; our interest is in the section $P(c) \cap T$. The special section $P(0) \cap T$ consists, as we know, of C_1 and C_2.

First, the sections $P(c) \cap T$ can be obtained in a simpleminded way by an empirical procedure. Take a wooden or rubber ring, or perhaps a small inflated inner tube, in the shape of our torus T, and suspend it by threads at the correct oblique angle over the surface of water. By carefully dipping T in water to a certain depth and then pulling it out, we can determine the wet boundary on T, which is obviously one of the sections $P(c) \cap T$. Then the dip is repeated to greater depth, and so on.

If it is wished to compute and plot the curves $P(c) \cap T$, their equation is obtained by putting $z_1 = c$ in (5) and solving the resulting equation for y_1 in terms of x_1:

$$y_1 = \pm \sqrt{a^2 + r^2 - c^2 - x_1{}^2 \pm 2a\sqrt{r^2 - (x_1 \sin \theta + c \cos \theta)^2}} \quad (13)$$

In plotting a curve from equation (13) it is necessary to limit oneself to values of x_1 that avoid square roots of negative numbers.

Two series of sections were obtained and plotted for the special torus with $a = \sqrt{2}$ and $r = 1$, for which the critical angle θ of equation (2) is 45°. Figure 4.7 is the exact analog of Fig. 4.3; it shows the sections of T by several planes inclined at 45° to the horizontal. For negative values of c the sections repeat themselves in reverse, much as they did in Fig. 4.3. The special values

$$c = 1 + \frac{1}{\sqrt{2}} = 1.7071 \qquad \text{and} \qquad c = 1 - \frac{1}{\sqrt{2}} = 0.2929. \quad (14)$$

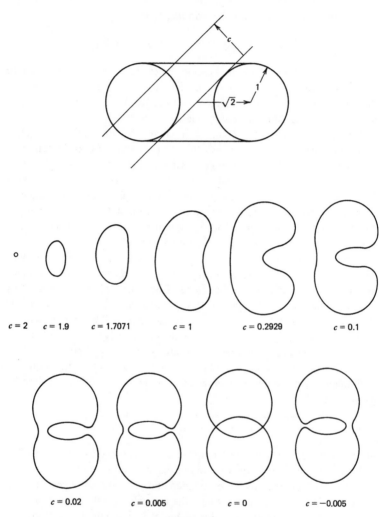

Figure 4.7

may be noted. The corresponding sections appear particularly flat, one from the right and the other from the left; they form transitions from convex to concave behavior. The torus T has the positively curved, or convex, outer part, and the negatively curved, or concave, inner part. The two regions are separated by the top and bottom horizontal circles

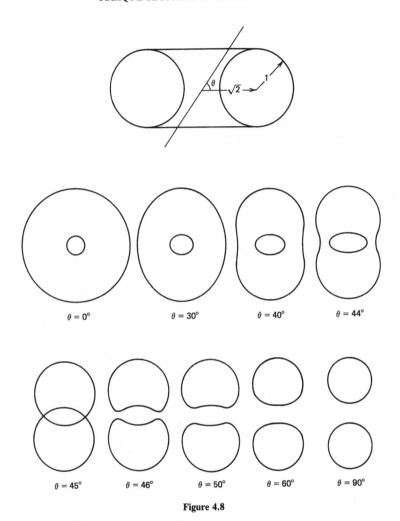

Figure 4.8

on T. The two critical values (14) correspond to planes tangent to those two circles. The particular flatness of the two sections comes therefore under curvature and inflexion phenomena.

Figure 4.8 shows several sections through the same torus by planes that pass through the origin, that is, have $c = 0$, of varying inclination to the horizontal. All four types of circles of the torus are shown here:

the horizontal circles C_H for $\theta = 0°$, the vertical circles for $\theta = 90°$, and the Villarceaux oblique circles for $\theta = 45°$.

EXERCISES

1. How many different face nets does a cube have?

2. If a convex p-gon is a plane section of a cube, what are the possible values of p?

3. A fly walks on a cubical box remaining equidistant from the end points of a body diagonal. What is its path?

4. A fly walks on a cubical box of unit size. Show that a path of length $\sqrt{5} + \epsilon$ will visit all the six faces of the box (ϵ is any small postive number) but no path of length $\sqrt{5}$ or less will do. (Hint: Consider the faces $ABEC$ and $CFHE$ in Fig. 4.4a).

5. Can you generalize the electrical network of Fig. 4.2b to a similar cubic network of 12 resistances, not all equal, but for which the same symmetry argument works?

6. The resistance of the electrical network of Fig. 4.2b is considered between (a) A and B, (b) A and G. All resistances are equal. Can both problems be answered by symmetry arguments?

7. When the square hole is punched through the unit cube Q and the biggest possible cube is pushed through the hole, the remainder of Q consists of four parts touching in points. What are these four parts?

8. The horizontal torus T has $a = \sqrt{2}$ and $r = 1$. The plane P is vertical and its distance from the center O of T is $\sqrt{2} - 1$. Compute and plot the sections of T by P and by two planes parallel to P and close to it from both sides.

9. Let T be the torus with parameters a and r. Let Z be the circular cylinder of base radius r and height $2\pi a$. Show without any integrations that the area of T is $4\pi^2 ar$ and its volume is $2\pi^2 ar^2$. (*Hint:* Slice T up into $2n$ "orange slices," rotate every second one by 180°, and let n increase; then compare with Z.)

CHAPTER FIVE

CONIC SECTIONS AND PASCAL'S THEOREM

The simplest curves of geometry are straight lines and circles. Then come ellipses, parabolas, and hyperbolas, collectively known as conic sections. These are usually studied by the method of analytic geometry: The ordinary rectangular coordinate system is introduced in the plane, a conic section is defined as the locus of all points with a certain property, and then an equation between x and y is developed that holds if and only if the point (x, y) has that property. In terms of the focal properties the definitions are as follows:

Ellipse: locus of all points in the plane for which the sum of distances from two fixed points (the foci) is constant

Hyperbola: locus of all points in the plane for which the difference of distances from two fixed points (the foci) is constant

Parabola: locus of all points in the plane equidistant from a fixed point (the focus) and a fixed line (the directrix).

The analytic method was introduced in the seventeenth century by the French philosopher and mathematician René Descartes. It is so powerful and far-reaching that it has largely overwhelmed the earlier synthetic method, which proceeds by purely geometrical constructions and without coordinates or algebra. However, the synthetic method was used in antiquity by the Greek geometers who discovered conic sections and developed a fairly complete theory. Their interest stemmed partly from attempts to solve the three classical problems of ancient geometry: to square the circle, to trisect any angle, to double the cube.

    The last task is known as the Delian problem because, according to a story, a priest of Apollo's shrine in Delos dreamt once that the god Apollo ordered his cubical altar to be doubled. According to the shrewd

Greek interpretation this amounted to constructing the side of a cube whose volume is twice that of a given cube. In simpler terms, what one wants is a geometrical construction of the number $\sqrt[3]{2}$. In the process of such a construction only the standard use of ruler and compasses is allowed. While $\sqrt{2}$ is easily obtained as the diagonal of the unit square, it turns out that $\sqrt[3]{2}$ cannot be so constructed at all. Thus the Greek attempts were doomed to failure but this was demonstrated by modern methods discovered some 22 centuries later.

It is lucky that the ancient Greeks did not know it—in their attempts they discovered things incomparably more valuable than geometrical recipes for cube doubling. In particular, they were thus led to the discovery of conics, for it was observed by them that finding $\sqrt[3]{2}$ can be reduced to intersecting two parabolas. To show this consider the parabolas

$$ay = x^2 \qquad \text{and} \qquad x = by^2,$$

which intersect in two points. One of them is the origin $(0, 0)$ and if the other one is (x_1, y_1) then

$$ay_1 = x_1{}^2, \qquad x_1 = by_1{}^2, \qquad y_1{}^3 = \frac{a}{b^2}.$$

Therefore with suitable a and b, for instance with $a = 2$ and $b = 1$, the cube root of 2 can be found.

It is a matter of considerable interest that conic sections, which two thousand years later were found by Kepler and Newton to describe the orbits of planets and the trajectories of comets in their courses about the sun, first appear because the curiosity of ancient Greeks was aroused. And according to the preceding story, that curiosity was aroused by, among other things, the dreams of priests.

In this section both the older synthetic and the newer analytic methods will be used and this not only for the sake of contrasting them. First, the synthetic method will be employed to show that ellipses, parabolas, and hyperbolas, as we have defined them, are in fact plane sections of circular cones. Briefly put, it is shown that conic sections are sections of cones.

An empirical demonstration of this is obtained by taking a lit flashlight with a reasonably accurately conical reflector. When the conical light beam is projected straight onto a wall, the area lit up is circular. As the flashlight is tilted away from the wall, this area becomes a more and more elongated ellipse. At a certain critical value of the tilt angle, when the outer edge of the beam becomes parallel to the wall, the ellipse opens out into a parabola. Still further tilting away from the wall will produce a hyperbola, or rather one branch of a hyperbola, as the shadow boundary.

It will be observed that the analytic method treats the plane conics purely inside the plane, in their own habitat, so to say. On the other hand, the synthetic method requires us to step out of the plane of the curve and into the three-dimensional space; we need to operate in *three* dimensions to find out what happens in *two*. Similar passage or, as it is called, embedding into a higher space, occurs in several branches of geometry, for instance, in projective geometry or in topology. Here this passage from plane to space continues the aim of the previous section: to help develop spatial visualization. Note that just as in the last section, so here too we are interested in the plane sections of a solid (or its surface).

Two final remarks will be made before the proofs themselves. Lest too much be ascribed to the ancient Greek genius, it must be added that our proofs are not exact copies of the Greek work. The latter was considerably more involved; the simplification and methods we use, though well within the scope and possibilities of Greek geometers, were missed by them. They are due to two nineteenth-century Belgian mathematicians, A. Quetelet (1796–1874) and G. Dandelin (1794–1847).

Our last point concerns the names *ellipse, hyperbola, parabola*. They derive from three Greek verbs meaning "to fall short of," "to throw further than," and "to throw as far as," and refer to the focus-directrix definition of conics that we take up a little later. The existence of three such highly specialized verbs may be explained as follows. Besides the sword, the ancient Greeks used spear and bow as their favorite weapons. Also, the main Olympic athletic event in ancient Greece was the pentathlon, a five-part contest in long jump, discus throw, javelin cast, running, and wrestling. Thus the names for the three conic sections in almost all modern languages may be ultimately due to Greek preferences in weapons and sports.

For the case of the ellipse we consider a circular cone C with vertex V, in the vertical position shown in Fig. 5.1. The intersecting plane P is tilted so that it cuts every generator of C. We recall that a generator is a straight line on C; it passes necessarily through the vertex V. It is easily seen that the curve E in which the plane P cuts the cone C is a bounded closed curve. Our object is to show that this E is an ellipse according to the definition given before. This requires that we produce in the plane P

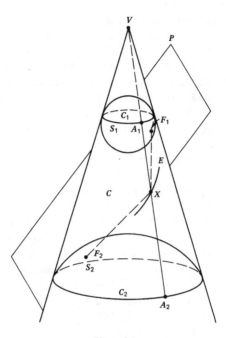

Figure 5.1

of E two points F_1 and F_2 which will function as the foci, and that we
show the sum

$$XF_1 + XF_2$$

to be the same for every point X of E.

On a first, superficial, look this might appear to run against our intuition. We know that an ellipse has two axes of symmetry at right angles,
and a center of symmetry that is at the intersection of those two axes.
This is particularly clearly brought out in the analytical treatment of the
ellipse. Its equation is

$$\frac{x^2}{a^2} + \frac{y^2}{b^2} = 1$$

and it is noted that only squared terms appear. Hence x may be validly
replaced by $-x$, and y by $-y$. These algebraic replacements are exact
counterparts of geometric symmetries.

On the other hand, it might appear that the section curve E of the cone C by the plane P cannot possibly have such symmetries. However, when the situation is examined more closely it will be seen that the hypothetical confusion of symmetries is due to the tacit and *incorrect* assumption that the center of the ellipse E must lie on the vertical axis of C—it does *not*.

The crucial points F_1 and F_2 are produced by the simple and ingenious Quetelet-Dandelin construction. Let S_1 be the sphere inscribed into the cone C and touching the plane P from above, say at the point F_1. Similarly, let S_2 be the sphere inscribed into the cone C and touching P from below, say at F_2. Let C_1 and C_2 be the horizontal circles of contact of the spheres with the cone, as shown in Fig. 5.1. We take an arbitrary point X of the section E of C by P. Consider now the generator of C that passes through X, that is, the straight line VX on C. Let it cross the circle C_1 at A_1 and the circle C_2 at A_2.

Since all the tangents drawn to a sphere from an external point are of equal length we have

$$XF_1 = XA_1 \quad \text{and} \quad XF_2 = XA_2.$$

Hence

$$XF_1 + XF_2 = XA_1 + XA_2 = A_1A_2.$$

But the quantity A_1A_2 is the distance between the circles C_1 and C_2 measured along *any* generator, and it is *independent* of the position of X on E. Therefore $XF_1 + XF_2$ is constant for all X so that E is an ellipse with the foci F_1 and F_2, q.e.d.

To obtain the parabola we keep our cone C as before and we align the plane P cutting C so that it is parallel to a generator of C. This is shown in Fig. 5.2 where the generator parallel to P is L, appearing to the left. Again, a sphere is inscribed into C, touching P from above, say at the point F. C_1 is the circle of contact of S with C, H is the horizontal plane containing the circle C_1, and the planes H, P intersect in the horizontal line D. Let Π be the curve of intersection of P with C. Let X be an arbitrary point on Π and let C_2 be the horizontal circle on the cone C through the point X. Finally, let the generator VX cut the circle C_1 at A.

We reason as before: All tangents to the sphere S from an external point F are of equal length. In particular, $XF = XA$. Here the distance XA between X and A is the distance between the circles C_1 and C_2 measured along *any* generator. In particular, we can choose to measure that distance along the generator L parallel to the plane P. With that choice

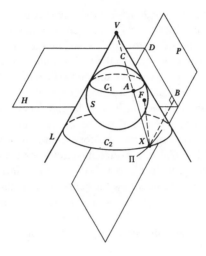

Figure 5.2

we find that XF, or XA, is the distance XB of X from the straight line D (since L is parallel to XB). Hence the distances of X to F and to D are equal. It follows that Π is the parabola with the focus F and the directrix D, q.e.d.

For the hyperbola we must take the complete *double* cone as shown in Fig. 5.3a. The plane P cuts both the upper and the lower parts of the cone. The reader is invited, and even urged, to visualize the geometry of the situation, *without* a drawing (i.e., without detailed drawings such as Figs. 5.1 or 5.2). For this purpose we might start with a vertical plane P. Various spheres, circles, and so on, will have subscripts 1 in the upper half of the cone, and 2 in the lower. We begin by inscribing a sphere S_1 into the upper part and a sphere S_2 into the lower part. Both spheres touch P, the upper one at F_1, the lower one at F_2. We also need the two circles of contact: C_1 where S_1 touches the cone, and C_2 where S_2 does.

The curve H of intersection of P with the double cone has two separate branches. Let us take any point X on the upper branch. As before, we use the fact that all tangents to a sphere are of equal length. In particular, XF_1 equals the distance from X to C_1 (measured along the generator through X); the same is true about XF_2. Hence the difference $XF_2 - XF_1$ is the distance between C_1 and C_2, measured along any generator, so it is constant. Therefore H is a hyperbola. We can now convince ourselves that everything works in the same way if the plane P is oblique so long as it cuts both halves of the cone. It may be surprising that even when P is tilted, the two branches of H are still symmetrically located.

We mention briefly an occurrence of the hyperbola familiar to most of us. When we take an ordinary nut of hexagonal shape, which fits over its bolt, then in front view we see something like Fig. 5.3b. What are the six congruent and repeating curvilinear arcs? When the manufacture of nuts is examined, it is found that they start from a horizontal slice of a hexagonal metal rod. This is drilled and threaded. To remove the sharp edges on the upper face, the slice is beveled down, by rotating it around its axis against an oblique file, for instance. This amounts geometrically to producing a conical cap (partial only) over the hexagonal slice. Therefore the six arcs are the intersections of a cone with the six vertical planes. Hence we have six identical hyperbolic segments.

A different description of conic sections will be given now, by means of the focus-directrix property. This is closer to the original Greek approach and has the advantage of providing a single common definition for ellipses, hyperbolas, and parabolas. However, as will be seen, circles will play a special role and will have to be excepted. Since the cone is always taken as being vertical, the circles are its plane horizontal sections.

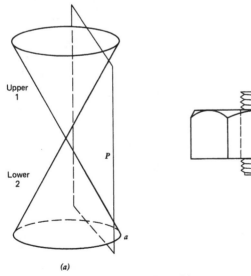

(a)

(b)

Figure 5.3

We start with a vertical cone C as in Fig. 5.4a intersected by an oblique plane P. Since sections of C by parallel planes are similar figures, the plane P could be moved up or down, with only scale changes occurring for the section. Thus the type of section is given by two things: the size of the aperture of the cone C and the steepness of the plane P. The aperture of the cone is conveniently measured by the semivertical angle α; this is the angle between the vertical axis of C and any generator of C, as is shown in Fig. 5.4a. The steepness of P may be given by the angle β defined as follows: Take any horizontal line L in P, and let L_1 be a line in P at right angles to L; then β is the angle L_1 makes with the vertical. Equivalently, β is the smallest angle a line in plane P makes with the vertical. It will be assumed that $0 \le \beta < 90°$, which prevents P from being

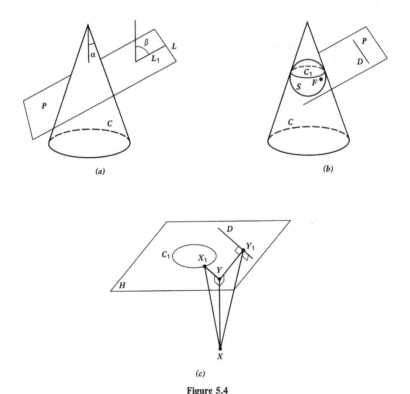

(a)

(b)

(c)

Figure 5.4

horizontal. The two angles α and β determine the nature of the plane section of C by P, and we get

An ellipse if $\beta > \alpha$,
A parabola if $\beta = \alpha$,
A hyperbola if $\beta < \alpha$.

The focus-directrix property is the following: A conic section is the locus of points in the plane for which the distance from a fixed point in the plane (the focus) has a constant ratio e to the distance from a fixed line in the plane (the directrix). The constant e is called the eccentricity of the conic, and we get an ellipse for $e < 1$, a parabola for $e = 1$, and a hyperbola for $e > 1$. It is therefore reasonable to expect that e is simply related to the angles α and β; in fact, it will be shown that

$$e = \frac{\cos \beta}{\cos \alpha} \tag{1}$$

Horizontal sections, the circles, are excepted by the condition $\beta < 90°$ so that $\cos \beta > 0$ and therefore $e > 0$. This suggests that circles are ellipses of eccentricity 0, which is the case.

We prove now the focus-directrix property. A plane P and a vertical cone C are taken in an arbitrary relative position, given by the angles α and β as in Fig. 5.4a. The first thing to do is to find the point F and the line D, in the plane P, which will function as the focus and the directrix of the conic curve that is the section of C by P. For this purpose we use again the Quetelet-Dandelin construction. A sphere S is inscribed into the cone C touching P from above at the point F, as is shown in Fig. 5.4b. This sphere S touches the cone C in a horizontal circle C_1; let H be the horizontal plane containing C_1 (this plane is not shown in Fig. 5.4b). Since P is not horizontal, the plane H cuts the plane P in a line D. It is not hard to show that the section curve in which P cuts C lies below H.

Consider now an arbitrary point X on the conic section $P \cap C$, and project it perpendicularly up onto the plane H as the point Y in H. Let the generator of the cone, passing through X, cut the horizontal circle C_1 at the point X_1 in H, as is shown in Fig. 5.4c. Let Y_1 be the point on D such that XY_1 is the distance of X to D, so that $XY_1 \perp D$. We obtain now two right-angled triangles XYX_1 and XYY_1 that lie in two vertical planes. Further, $\sphericalangle X_1 XY = \alpha$ and $\sphericalangle YXY_1 = \beta$. Now consider the ratio FX/Y_1X, which is the ratio of distances of X to F and to D. Since XX_1 and FX are two tangents to the sphere S from X, they are equal; therefore

$$\frac{FX}{Y_1X} = \frac{X_1X}{Y_1X}.$$

But $XX_1 = XY \sec \alpha$ and $XY_1 = XY \sec \beta$ because the two right-angled triangles share the side XY. Hence

$$\frac{FX}{Y_1X} = \frac{XY \sec \alpha}{XY \sec \beta} = \frac{\cos \beta}{\cos \alpha},$$

which is the constant eccentricity e. This proves the focus-directrix property: The conic section is the locus of points X in the plane such that the ratio of distances from X to F and to D is the constant e given by (1).

A consequence of the focus-directrix property of the conics is the very convenient *single* equation for all of them in polar coordinates. Let p be the distance of the focus F of the conic from its directrix D, as is shown in Fig. 5.5a, and let X be any point on the conic; θ and r are the usual polar coordinates of X. Then r is the distance of X to F, and $p + r \cos \theta$ is its distance to D. Therefore by the focus-directrix property

$$\frac{r}{p + r \cos \theta} = e.$$

Multiplying out, rearranging, and putting $b = pe$, we get

$$r = \frac{b}{1 - e \cos \theta} \tag{2}$$

as the polar equation of the conic.

It is an easy matter now to prove that the path of a celestial body,

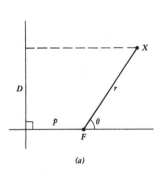

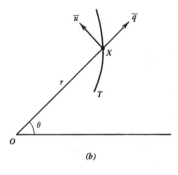

(a)

(b)

Figure 5.5

planet or comet, moving under the inverse-square law of attraction by the sun is a conic for which the sun is a focus. We say, "It is an easy matter now to prove . . ." and we shall prove it, but it must be remembered that the matter was not always easy, and that it was for *this* purpose that Newton developed the apparatus known as calculus.

It is reasonable to describe planetary motion in polar rather than Cartesian coordinates, and we start by getting the radial and transverse components of acceleration. Let $r = r(t)$ and $\theta = \theta(t)$ be the polar coordinates of a point X moving along trajectory T, expressed as functions of time t. Let $\bar{q}$ and $\bar{u}$ be the unit vectors in the radial and transverse directions, shown in Fig. 5.5b. These vectors are also functions of time t and are moving along with X; their components are

$$\bar{q} = (\cos \theta, \sin \theta), \qquad \bar{u} = (-\sin \theta, \cos \theta) \qquad (3)$$

In honor of Newton's fluxions differentiation with respect to time t is denoted by placing a dot above the quantity differentiated. Using (3) and the chain rule we have

$$\dot{\bar{q}} = \dot{\theta}\bar{u}, \qquad \dot{\bar{u}} = -\dot{\theta}\bar{q} \qquad (4)$$

Let $\bar{x}$ be the displacement vector giving the position of X:

$$\bar{x} = r\bar{q}.$$

Differentiating this with respect to t, using the product rule, and (4), we get the velocity vector

$$\bar{v} = \dot{r}\bar{q} + r\dot{\theta}\bar{u}$$

and differentiating again we obtain the acceleration vector

$$\bar{a} = (\ddot{r} - r\dot{\theta}^2)\bar{q} + (r\ddot{\theta} + 2\dot{r}\dot{\theta})\bar{u}, \qquad (5)$$

which is the formula we need.

Suppose now that O represents the sun and X the moving planet or comet and that the only force acting on X is the Newtonian attraction straight toward O. Since this force is radial the total acceleration is also radial and so the tangential or transverse $\bar{u}$ component in (5) must vanish:

$$r\ddot{\theta} + 2\dot{r}\dot{\theta} = 0.$$

But

$$r\ddot{\theta} + 2\dot{r}\dot{\theta} = \frac{1}{r}(r^2\dot{\theta})^{\cdot};$$

therefore

$$r^2\dot{\theta} = c \tag{6}$$

is a constant. This, incidentally, is the second law of Kepler, stating that the radius joining the sun to the planet sweeps out area at a constant rate. We recognize this at once by recalling the area formula in polar coordinates

$$A = \frac{1}{2} \int r^2 \, d\theta,$$

which shows that $\dot{A} = r^2\dot{\theta}/2$. Therefore (5) reduces to

$$\bar{a} = (\ddot{r} - r\dot{\theta}^2)\bar{q}.$$

Here we take magnitudes, apply Newton's law that force is mass times acceleration, and use the form of Newton's law of gravitational attraction, getting

$$\ddot{r} - r\dot{\theta}^2 = \frac{-K}{r^2} \tag{7}$$

Here K is a positive constant and the minus sign arises since gravity pulls. Next,

$$\frac{d(1/r)}{d\theta} = \frac{d(1/r)}{dt} \Big/ \frac{d\theta}{dt} = \frac{-\dot{r}}{r^2\dot{\theta}},$$

so that by (6)

$$\frac{d(1/r)}{d\theta} = \frac{-\dot{r}}{c}.$$

Therefore

$$\frac{d^2(1/r)}{d\theta^2} = \frac{d}{d\theta}\left[\frac{d(1/r)}{d\theta}\right] = \frac{-\ddot{r}}{c\dot{\theta}}$$

so that

$$\ddot{r} = -c\dot{\theta}\frac{d^2(1/r)}{d\theta^2}.$$

Using this and (6) we find that (7) becomes

$$\frac{d^2(1/r)}{d\theta^2} + \frac{1}{r} = C_1,$$

where $C_1 = K/c^2$. Putting $y = 1/r$, the preceding expression becomes

$$\frac{d^2y}{d\theta^2} + y = C_1.$$

This is a simple nonhomogeneous differential equation for which the general solution may be written as

$$y = C_1 + C_2 \cos \theta + C_3 \sin \theta \tag{8}$$

with two arbitrary constants C_2 and C_3. Equivalently, we introduce the more suitable two constants: the amplitude A and the phase θ_0, and (8) then becomes

$$y = C_1 - A \cos(\theta - \theta_0).$$

Returning to the original r, the radius joining the moving planet or comet to the sun, we find that

$$\frac{1}{r} = C_1 - A \cos(\theta - \theta_0),$$

so that

$$r = \frac{1/C_1}{1 - (A/C_1) \cos(\theta - \theta_0)}.$$

Comparison with (2) shows now that the trajectory is indeed a conic.

We have traced some outlines of the development of conic sections by the synthetic method of ancient Greek geometry. The real flowering of the analytic theory of the conics was in connection with Kepler's laws of planetary motion and Newton's work in the seventeenth century. The geometric flowering occurred somewhat earlier.

Thirteen centuries separate Pappus of Alexandria, third century A.D., the last of the great Greek geometers of antiquity, from the outstanding French geometers Descartes (1596–1650) and Pascal (1623–1662), and not very much was done in geometry in that time. However, once the new spirit and the new techniques came in, progress was swift and immense. We shall now prove a theorem discovered by the sixteen-year-old Blaise Pascal in 1639: If a hexagon is inscribed into a conic, then the pairs of its opposite sides intersect in three points that are collinear.

This theorem is quite easy to state, its conclusion is very unobvious, and, what is perhaps most surprising, its consequences are absolutely fundamental, certainly far more so than might be expected from such a seeming oddity. Pascal himself deduced from it over 400 consequences, including, besides much else, a large part of the work of Apollonius, the greatest ancient Greek authority on the conic sections.

The usual proof of Pascal's theorem offered nowadays is by an appeal to projective transformations and projective geometry. This reduces the general case to a case so special as to be obvious. There is some analogy here to our proof of Routh's theorems where the case of an arbitrary triangle is deduced from the special case of a right-angled or an equilateral triangle. However, proving Pascal's theorem by projective geometry is in a way like putting the cart before the horse: projective geometry did not give rise to Pascal's theorem; it was the other way around.

Besides, projective geometry is not introduced here, and so another proof must be given. We shall offer a modification of Pascal's original proof which is brief and self-contained and whose essence is analytical and Cartesian, in complete distinction to the Greek synthetic method used earlier in this section. The idea is to exploit the Cartesian method of handling conics C and lines L by means of their equations $C = 0$ or $L = 0$. Here $C(x, y)$ is a quadratic polynomial and $L(x, y)$ is a linear polynomial. Intersection properties are heavily used and we start with them.

Let $C_1 = 0$ and $C_2 = 0$ be two conics and let k be a constant. Then $C_1 + kC_2 = 0$ is also a conic, passing through the intersection points of C_1 with C_2. This is simply proved: The new equation represents a conic, since in it a quadratic polynomial $C_1 + kC_2$ is set equal to zero, and the new conic passes through every point common to C_1 and to C_2 because the coordinates of such a common point satisfy both equations $C_1 = 0$ and $C_2 = 0$, hence also $C_1 + kC_2 = 0$. Conversely, any conic through the intersection points of C_1 and C_2 has the equation $C_1 + kC_2 = 0$ for some constant k. The sole exception is C_2 itself, and even this exception is removed if $C_1 + kC_2 = 0$ is replaced by the homogeneous form $k_1C_1 + k_2C_2 = 0$.

An important use will also be made of the degenerate hyperbola consisting of two intersecting lines. Such two lines do form an honest though degenerate hyperbola: a conic section in which a cone is cut by a plane through the vertex. In terms of the focus-directrix property the eccentricity of such a degenerate hyperbola is $e > 1$ but the focus lies on the directrix. Hence the polar equation (2) fails; however, the Cartesian equation holds: In the general equation

$$\frac{x^2}{a^2} - \frac{y^2}{b^2} - k = 0,$$

we put $k = 0$.

Before proving Pascal's theorem itself we prove a preliminary result. Let $C = 0$ be a conic and $L = 0$ a straight line cutting that conic, as shown in Fig. 5.6. Now a conic C_1 through the intersection points of C with L has the equation $C + k_1LL_1 = 0$, where k_1 is a constant and L_1 is some first-degree polynomial in x and y. We take two such conics C_1 and C_2, shown in Fig. 5.6, and with the equations

$$C + k_1LL_1 = 0, \qquad C + k_2LL_2 = 0.$$

Every two of the three conics C, C_1, C_2 intersect in four points. Two of such four points lie by construction on the line L but the other two determine a new straight line. We get thus three straight lines, shown dotted in Fig. 5.6, with the equations

$$L_1 = 0, \qquad L_2 = 0, \qquad k_1L_1 - k_2L_2 = 0.$$

It follows that the three lines pass through a point (P in Fig. 5.6).

Now for Pascal's theorem itself. Let a, b, c, d, e, f be the consecutive vertices of a hexagon H, which lie on a conic C. This is illustrated in Fig. 5.7, where C is taken as a hyperbola. We set up the situation of our preliminary result as follows:

conic C	is	C
line L	is	ad
conic C_1	is	ab, cd
conic C_2	is	af, ed

That is, the conics C_1 and C_2 are degenerate hyperbolas, i.e., line pairs. Now, applying the preliminary result we get Pascal's theorem at once.

To show this at some length, the three pairs of lines are traced as

Pair of conics	Leads to line
C, C_1	bc
C, C_2	fe
C_1, C_2	BA

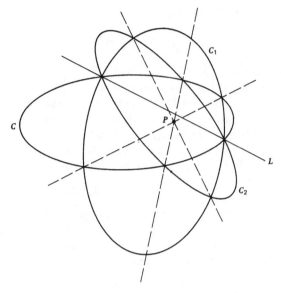

Figure 5.6

So, as is shown in Fig. 5.7, the point D in which bc cuts fe lies on BA. This is exactly the conclusion of Pascal's theorem, and DBA is the Pascal line P.L. for the hexagon $abcdef$ inscribed into the conic C.

The hexagon of Pascal's theorem is allowed to intersect itself; two examples of self-crossing hexagons are shown in Fig. 5.8 together with their Pascal lines. The conic C is an ellipse in one case and a parabola in the other.

The next observation is that the conic C into which the hexagon H is inscribed in Pascal's theorem may itself be degenerate, that is, may be a pair of intersecting lines. Then Pascal's theorem gives us, as a special case, the theorem of Pappus: If in a self-crossing hexagon H the vertices lie alternately on a pair of lines L_1 and L_2, then the opposite sides of H cut in three points that lie on a straight line L. This line L may be called the Pappus line for the hexagon H on L_1 and L_2. There are two possibilities for the type of self-crossing of H, shown in Fig. 5.9. Actually each case in Fig. 5.9 provides not one but three applications of Pappus' theorem: there is a hexagon on every two lines, with the third one as the Pappus line.

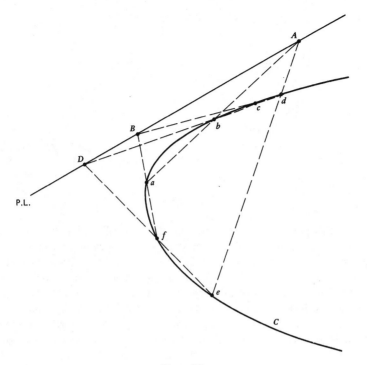

Figure 5.7

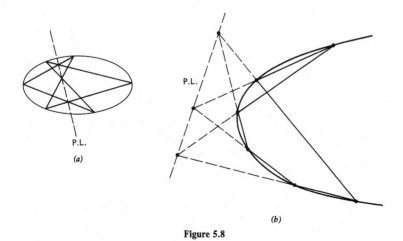

Figure 5.8

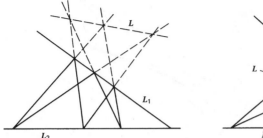

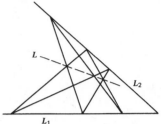

Figure 5.9

Another possibility is when the conic C of Pascal's theorem is a circle. In this special form Pascal's theorem has some applications to elementary geometry; we mention one very briefly. Let ABC be a triangle of Fig. 5.10 and P any point in it. Project P at right angles on the two sides AC and BC; let P_1 and P_2 be the projections. Draw the transversals AP and DP and project the vertex C on these two transversals; let Q_1 and Q_2 be the projections. Then the lines Q_1P_1 and Q_2P_2 intersect on AB. The proof follows at once by applying Pascal's theorem to the *crossed* hexagon $CP_1Q_1PQ_2P_2$. The conic is here the circle on CP as diameter and AB is the Pascal line.

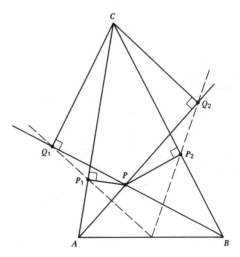

Figure 5.10

Let us consider now the applications of Pascal's theorem to conics proper. Here very important results are obtained when two consecutive vertices *a* and *b* of the hexagon *H* are allowed to merge. As limit considerations suggest, when *b* coalesces with *a*, then the side *ab* of *H* should be replaced by the tangent *T* to the conic at *a*. It is important to realize here that Pascal worked before Newton's discovery of calculus, which has trivialized the central problem of finding tangents to curves. We have seen once already, in the work of Descartes on rolling, in Chapter 3, how geometry grappled with that problem.

The five-point version of Pascal's theorem has the following consequence: it allows us to construct, *with ruler alone*, the tangent *T* to a conic *C* at a point *a* on *C*, if four other points *c, d, e, f* of *C* are given. The construction is illustrated in Fig. 5.11 with a pentagon *acdef* inscribed into an ellipse. If that pentagon is regarded as a degenerate hexagon *aacdef,* then Pascal's theorem tells us that the following three pairs of straight lines intersect in points that are collinear: *ac* with *ef* at *A*, *cd* with *af* at *B*, tangent *T*, or *aa* with *de* at *X; A, B, X* collinear. The points *A* and *B* are obtained by ruler alone and so is the Pascal line *AB*. Next *X* is found as the intersection of the lines *AB* and *de*. Finally, the desired tangent *T* is obtained as the line *aX*.

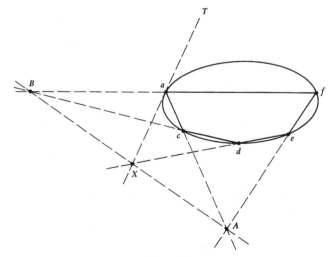

Figure 5.11

A similar result is obtained from the three-point case of Pascal's theorem. Let a triangle $p_1p_2p_3$ be inscribed into a conic C and let the tangents to C at p_1, p_2, p_3 be T_1, T_2, T_3. Regarding the triangle $p_1p_2p_3$ as the degenerate hexagon $p_1p_1p_2p_2p_3p_3$, we express Pascal's theorem as follows: Let each side of the triangle intersect the tangent at the opposite vertex; then the three points of intersection lie on a line. It follows that if three points are known on a conic C as well as the tangents to C at two of them, then the third tangent can be obtained by ruler alone.

An especially important consequence of Pascal's theorem is the exact number of points necessary to determine a conic. As an analogy we know that two points determine a line and three points determine a circle. Actually it is necessary to modify this to the following: Two distinct points determine a line, three noncollinear points determine a circle. One sometimes puts it as: Two points in general determine a line, three points in general determine a circle. How many points in general determine a conic? By using methods developed since Pascal's day we can easily answer a more general question. A conic is a quadratic, or second-degree, curve $C(x, y) = 0$ where $C(x, y)$ is a polynomial of degree two in x and y. When $C(x, y)$ is a polynomial of degree n, the corresponding curve is called an nth degree curve. How many points in general determine an nth degree curve?

The nth degree polynomial $C(x, y)$ has the general form

$$A_1x^n + A_2x^{n-1}y + \cdots + A_{N-1}y + A_N = 0 \qquad (9)$$

and we find first the number N of different terms. There are $n + 1$ of them of degree n; these go with the $n + 1$ possible power combinations

$$x^n, x^{n-1}y, \ldots, xy^{n-1}, y^n$$

Similarly, there are n terms of degree $n - 1$, $n - 1$ terms of degree $n - 2$, and so on, down to two terms of degree 1, for x and for y, and one constant term of degree 0. The number of all terms in (9) is therefore

$$1 + 2 + 3 + \cdots + n + (n + 1) = \frac{(n + 1)(n + 2)}{2}$$

If the locus of (9) is to be of any interest, not all coefficients A_i vanish—we divide (9) by any one $A_i \neq 0$. This leaves us with

$$\frac{(n + 1)(n + 2)}{2} - 1 = \frac{n(n + 3)}{2}$$

unknown coefficients A_i to be found. How many points must be given to fix the curve? Each point gives us one linear equation in the unknown coefficients; therefore the number of points must equal the number of coefficients: the number of points necessary to determine a general nth degree curve is $n(n + 3)/2$. For $n = 1$ we get 2 and indeed two points determine a line. For a conic $n = 2$ and so five points are in general necessary.

This will be proved now as a consequence of Pascal's theorem: If a, b, c, d, e are points on a conic C, then C is the only conic passing through them. Let f be a sixth point on C whose position we do not yet know. Then in the hexagon $H = abcdef$ only one pair of opposite sides is known to us; which one it is will depend on the position of f relative to a, b, c, d, e on C. For instance, if f lies between a and e, then the known pair is ab and de. This known pair intersects in a point A that can be found, and the Pascal line for H must pass through A. Take now any line L through A as the Pascal line for H. Then, working backward, we can find the point f. In the case given above we let bc cut L at M and cd cut L at K. Then f is the intersection of aK and eM. The sixth point f depends on the line chosen as L, and as L rotates about its fixed point A the sixth point f traces out the conic C. Hence C is determined by its five points a, b, c, d, e.

One final matter must be discussed. It was tacitly assumed that the opposite pairs of sides of the hexagon H intersect when we want them to intersect. But what if they do not, being parallel? It is here that the projective formulation from the end of Chapter 2 comes into force. It will be recalled that every line has one point at infinity, any two parallel lines share their point at infinity, and there is one line at infinity for the whole plane containing the infinite points of all lines in the plane. Consider now the case of Pascal's theorem illustrated in Fig. 5.12a. Here one pair of opposite sides, ab and de, are parallel, bc cuts ef at B, and cd cuts af at A. The "intersection" for the parallel pair, ab and de, is their point at infinity; it follows now that this point at infinity must lie on AB by Pascal's theorem. Therefore the Pascal line AB is parallel to ab and to cd.

Consider next the case of Fig. 5.12b. Here ab and de are parallel, and so are bc and ef. In other words, the Pascal line for the hexagon $abcdef$ is the line at infinity. Therefore the third pair of opposite sides, cd and af, must also be parallel. The conclusion of Pascal's theorem holds: Pairs of opposite sides intersect in collinear points. However, the three inter-

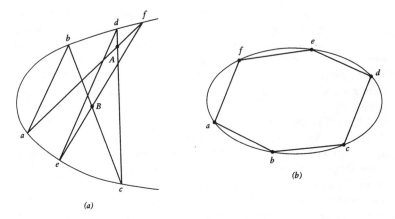

(a)

Figure 5.12

section points are points at infinity, and the Pascal line is the line at infinity.

These, and similar, illustrations show how with the projective formulation we avoid special or exceptional cases.

EXERCISES

1. Derive the Cartesian equation of the ellipse, $x^2/a^2 + y^2/b^2 = 1$, from the focal property. (*Hint:* Choose origin and axes cleverly.)

2. Find the locus of all points in the plane equidistant from a circle and a point inside that circle.

3. Do the same when the point lies outside the circle.

4. Show that on any fixed circular cone any ellipse is obtained as a suitable plane section.

5. Show that the same is not true for hyperbolas. Which hyperbolas can be obtained as plane sections of a fixed cone?

6. A circular cone C is given and A is a point inside. How many plane sections, if any, of the cone are ellipses with a focus at A?

7. What is the locus of points in the plane of an ellipse E, from which the two tangents to E are perpendicular?

8. Show that one ellipse can touch as many as four distinct concentric circles.

9. An antiparallelogram is a self-crossing quadrilateral $ABCD$ in which AB cuts CD while $AB = CD$ and $AD = BC$. Let A and D stay fixed while B and C move so that all sides keep their fixed length. On what curve does the crossover point of AB with CD move?

10. A diameter of an ellipse is defined to be the locus of midpoints of any family of parallel chords of the ellipse. Show that it is a straight segment passing through the center of the ellipse. (*Hint:* The ellipse is an orthogonal projection of a circle.)

11. An ellipse is drawn on an otherwise blank page. How would you find its center?

12. In an ellipse E all chords are drawn that cut off a segment of some constant area. Show that the chords are tangent to a similar ellipse with the same center as E. (*Hint:* Treat E as the orthogonal projection of a circle K; what happens to the chords in K that project onto the chords of E cutting off a fixed area?)

CHAPTER SIX
EXAMPLES OF
GEOMETRICAL EXTREMA

In this section we collect and solve a number of problems in which a geometrically given quantity, such as a length, an area, or a volume, is to be made largest or smallest. The intention is (1) to illustrate and supplement standard techniques of calculus; (2) to show the benefit of physical, especially mechanical, analogs; (3) to exhibit the occurrence and importance of boundary extrema; and (4) to introduce certain new geometrical ideas. Several such problems on geometrical extrema have been handled already—for instance, the isoperimetric problem for polygons in Chapter 1 and the longest-rod problem in Chapter 3.

Problem 1. Find the rectangle(s) R of largest area that can be inscribed into a given acute-angled triangle T.

Here one exploits first a certain rigidity of the extremal rectangle R, showing that its four vertices lie on T and not inside T. If only 0, 1, or 2 vertices of R lie on T, then it is clear that R can be moved into T and so its area is not maximal. But the case of three vertices on T needs a special consideration. Suppose then that we have the configuration of Fig. 6.1a. It will be shown that R can be rotated by a small angle α about some center O so as to fall inside T. Choose the sense of rotation shown by the arrow. Where could O possibly lie? Let L be the line through the vertex c of R at right angles to the side of T. If c is to be rotated *into* T, then O must lie in the half-plane H_c consisting of all points on the same side of L as b or on the dotted part of L. In the same way we produce H_b and H_a.

The center O must therefore lie in H_a, in H_b, and in H_c; it is a point of their intersection, or common part, $S = H_a \cap H_b \cap H_c$. But must S always contain at least one point? No, it could be empty. In fact, there are three possible cases shown in Fig. 6.1b. There is no difficulty with the top or middle case (see problem 1 in the Exercises). If the bottom

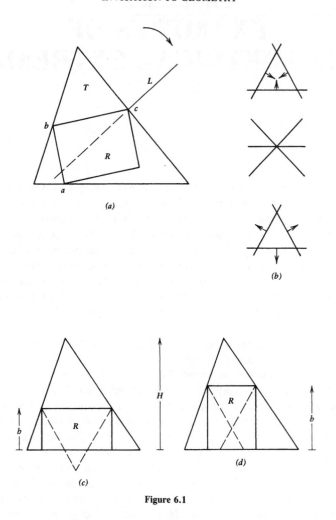

Figure 6.1

case occurs, reverse the sense of rotation to its opposite; this changes each half-plane to its opposite and one has the top case again. As a bonus we get the exact region of the admissible rotation centers O.

All four vertices of R must lie on T, and so, since T has three sides, two vertices lie on a side. How high is R? Suppose its height h to be less than half the corresponding height H of T, as in Fig. 6.1c. Then the three residual triangles of T, left uncovered by R, can be bent over the sides of R so as to cover R, and with something left over. Hence here area R $< \frac{1}{2}$ area T. The same can be done if $h > H/2$ as in Fig. 6.1d. Therefore the maximum area of R is half that of T, and this maximum occurs if the rectangle stands on any one side of T as base and has the midpoints of the other two sides as vertices. If T is allowed to be obtuse, then there is only one solution instead of three: Only the side opposite the obtuse angle can contain the base of R.

Problem 2. Two circles pass through a point P as in Fig. 6.2; find the straight line through P so that the total intercept AB is longest.

Let O_1 and O_2 be the centers of the two circles and draw the circle on $O_1 O_2$ as diameter; let L be arbitrary. On this third circle let D be a point such that $AB \parallel DO_2$. Project P perpendicularly onto DO_2 as Q. Then

$$AP = 2DQ, \qquad PB = 2QO_2, \qquad AB = 2DO_2.$$

But DO_2 is a chord in the third circle; it is longest when the chord is the diameter. Hence the length AB is maximum when $AB \parallel O_1 O_2$, or, equivalently, when AB is perpendicular to the common chord of the two given circles.

Problem 3. Find the triangle $T = A_1 B_1 C_1$ of largest perimeter whose sides pass through three given points A, B, C and whose angles are given: α opposite A at A_1, β opposite B at B_1, γ opposite C at C_1 as shown in Fig. 6.3a.

Somewhat imprecisely restated, the problem is to find the triangle T of maximum perimeter that circumscribes one given triangle ABC and is similar to another given triangle. The positions of A_1, B_1, C_1 are unknown, but these points must lie on certain circles because of Thales' theorem. For instance, BC is to be seen at the angle α from A_1; hence A_1 lies on a circle K_1 through B and C, centered at O_1, which can be constructed, since B, C, and α are known. Similarly, we produce the circles K_2 and K_3; all three circles are shown in Fig. 6.3b. K_1, K_2, K_3 pass through a point M: Suppose that K_1 and K_2 cut at M; then

$$\sphericalangle AMC = 180° - \beta, \qquad \sphericalangle CMB = 180° - \alpha,$$

$$\sphericalangle AMB = 360° - \sphericalangle AMC - \sphericalangle CMB = \alpha + \beta = 180° - \gamma.$$

Therefore K_3 also passes through M (its chord AB is seen at the angle γ from C and at the angle $180° - \gamma$ from M). This proof applies to the case when M lies inside ABC; a similar proof applies to the outside case.

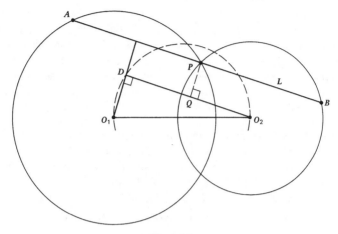

Figure 6.2

We have just proved a proposition sometimes known as Miquel's theorem, after a nineteenth-century Spanish mathematician A. Miquel: If three points lie one each on the sides of a triangle, and if the circle is drawn through each vertex and the two points on the sides from that vertex, then all three circles pass through one point, the Miquel point M. The proposition still holds even when the three points lie on the *extended* sides of the triangle.

We know therefore the locus of the three vertices A_1, B_1, C_1 of T: they must lie on certain arcs of the three circles K_1, K_2, K_3. The precise extent of these arcs is determined by the lines AB, AC, BC and by the tangents to the circles at A, B, C. Any one vertex A_1, B_1, C_1 can be chosen at will on its arc. For instance, if we choose A_1 on its allowed arc of K_1, then the lines A_1C and A_1B cut the circles K_2 and K_3 at B_1 and C_1, and B_1C_1 passes automatically through A. To finish the problem we now apply Problem 2:

$$A_1B_1 \leq 2O_1O_2, \qquad B_1C_1 \leq 2O_2O_3, \qquad A_1C_1 \leq 2O_1O_3.$$

Hence the maximum perimeter of $A_1B_1C_1$ is

$$2(O_1O_2 + O_1O_3 + O_2O_3)$$

and it is attained when the sides of $A_1B_1C_1$ are parallel to those of $O_1O_2O_3$. In other words, the maximum-perimeter triangle $A_1B_1C_1$ is similar to the known triangle $O_1O_2O_3$ in the ratio $2:1$ and with the Miquel point M as the center of similarity, as is shown in Fig. 6.3c. Put in yet a different way, the maximal triangle is obtained by drawing lines through A, B, C perpendicular to the chords AM, BM, CM. Since all triangles $A_1B_1C_1$ are similar, on account of having the same angles, it follows that the triangle that maximizes the perimeter also maximizes the *area*.

The next two problems exhibit the technique of reflection, which is often useful in minimizing distances. We start with a simple problem which is the basis of that technique:

Problem 4. Two points A and B lie on the same side of a straight line L. Find a point P on L so as to minimize the sum of distances $AP + BP$.

Let P be the minimizing point on L and reflect one of the two given points, say B, in the line L. As is shown in Fig. 6.4a, let B_1 be the reflection; we say that B reflects in L to B_1. Then $BB_1 \perp L$ and $BK = B_1K$ so that $PB = PB_1$. Therefore the quantity to be minimized is $AP + PB_1$; hence the points A, P, B_1 are collinear. It follows that AP and PB make equal angles with L: $\alpha = \beta$.

A theorem on the ellipse follows from the preceding. Let E be the ellipse with A and B as foci, passing through P. By the definition of an ellipse

$$AX + BX < AP + BP \qquad \text{for a point } X \text{ inside } E,$$

$$AX + BX = AP + BP \qquad \text{for any point } X \text{ on } E,$$

$$AX + BX > AP + BP \qquad \text{for any point } X \text{ outside } E.$$

It follows that the ellipse E is tangent to L at P, for otherwise we could move P on L into the interior of E, thus diminishing $AP + BP$. Since the angles α and β are equal, the following has been proved: A tangent to the ellipse makes equal angles with the straight segments joining the point of tangency to the foci. This is known as the focusing property of the ellipse, since it can be interpreted as follows: The rays of light from a focus of the ellipse, after reflecting internally once, focus again at the other focus of the ellipse.

We can keep one focus of an ellipse E fixed and move the other one to infinity. In terms of E as a section of a cone this means that the plane P that cuts the cone moves to a position parallel to a generator of the cone. In the limit one gets the focusing property of the parabola: Rays of light coming from the focus of the parabola form after the reflection a parallel beam. This is the basis of design of certain optical reflectors

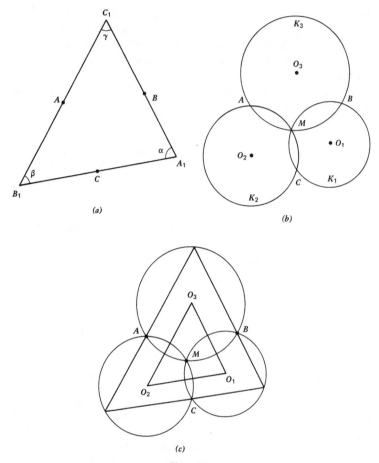

Figure 6.3

and radio antennas. The hyperbola also has a focusing property, that of virtual focusing: If rays of light fall on the outside of one branch of the hyperbola from the focus F_1 of the other branch, then after the reflection the rays appear to emerge from the other focus F_2.

Something of the focusing properties of the conics shows itself in the etymology of the word: *Focus* means "fireplace," or "hearth," in Latin, and by extension the central point of home, center of domestic worship.

The technique of simple reflection, shown in Fig. 6.4*a*, extends to multiple reflection. Suppose, for instance, that two straight lines L_1 and L_2 intersect at O, as in Fig. 6.4b, and A, B are two points in the angle L_1OL_2. We want to find points P on L_1 and Q on L_2 so as to minimize the sum $AP + PQ + QB$. Let the minimum occur for the points P, Q shown in Fig. 6.4b and reflect A in L_1 into A_1. By the previously proved result A_1, P, Q are collinear. Reflect the image A_1 in L_2 into A_2; then A_2, Q, B are collinear. Hence Q is the intersection of L_2 with BA_2, P is the

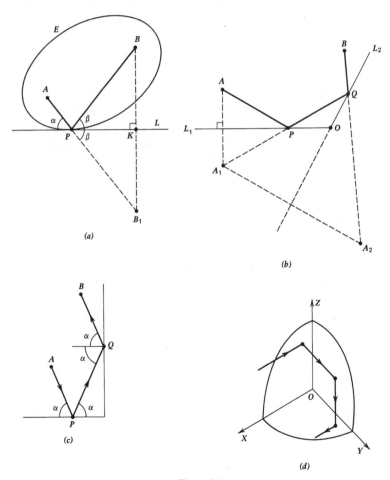

Figure 6.4

intersection of A_1Q with L_1, and our problem is solved. The same method obviously extends to any number of reflections.

Since rays of light move along shortest paths, the preceding may be interpreted optically: If L_1 and L_2 are mirrors, then the ray of light from A toward P hits B after two reflections. The same is true about a perfectly elastic billiard ball from A: If struck so as to move toward P, it bounces off the two banks and hits B.

An interesting special case arises when L_1 and L_2 are at right angles, as in Fig. 6.4c. Here it is found that *any* ray AP emerges after two reflections as QB, parallel to its original path AP. This is the principle behind the corner reflectors of Fig. 6.4d, used in the red warning lights on bicycles and cars and also in radio communication. Let three plane mirrors XOY, YOZ, ZOX be pairwise perpendicular. Let a ray of light arrive at one of the three mirrors; then after three reflections, once from each mirror, the reflected ray emerges parallel to its original path. The four segments forming the path of the ray lie in one oblique plane; let us suppose that $\bar{n} = (a, b, c)$ is the unit vector along the first segment. Then the unit vector along the second segment is $(a, -b, c)$, along the third segment it is $(-a, -b, c)$, and along the emergent ray it is $(-a, -b, -c)$. Thus after three reflections the ray is parallel to the original course.

Problem 5. Inscribe into a given acute-angled triangle ABC another triangle $A_1B_1C_1$ of the smallest perimeter.

This is known as Fagnano's problem, after the Italian mathematician marquess Gianfrancesco di Fagnano (1715–1797). Let the triangle $A_1B_1C_1$ be inscribed into ABC, as in Fig. 6.5a. Varying one vertex of $A_1B_1C_1$ at a time and applying Problem 4, we find that the sides of $A_1B_1C_1$ are inclined to the sides of ABC at equal angles, as is shown in Fig. 6.5a. Reflect C_1 in AC into C_2, and in BC into C_3. Then the perimeter of $A_1B_1C_1$, which is to be minimum, is equal to $C_2B_1 + B_1A_1 + A_1C_3$. Hence the four points C_2, B_1, A_1, C_3 are collinear. Also, our reflection of C_1 into C_2 and C_3 shows that

$$C_2C = C_1C = C_3C \quad \text{and} \quad \sphericalangle C_2CC_3 = 2\gamma.$$

Therefore

$$A_1B_1 + B_1C_1 + C_1A_1 = C_2C_3 = 2CC_1 \sin \gamma.$$

This is a convenient expression for the perimeter of $A_1B_1C_1$ since it depends on only one variable, the length CC_1. The minimum is obviously

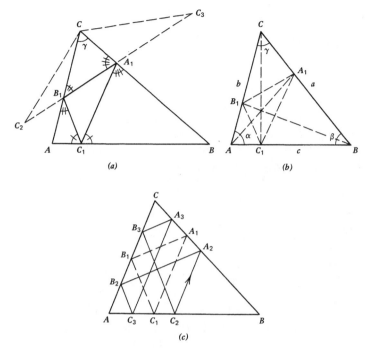

Figure 6.5

attained when C_1 is the orthogonal projection of C onto AB. By the same reasoning B_1 and A_1 are the projections of B and A onto the opposite sides. Hence the minimum-perimeter triangle is formed by the feet of the three altitudes of ABC; it is called the orthic triangle of ABC.

The six elements of the orthic triangle, that is, its three sides and three angles, are easily found. Expressing C_1B and A_1B in Fig. 6.5b in terms of a, b, c, α, β, γ, and applying the law of cosines to A_1C_1B and to the analogous two other triangles, we find

$$A_1C_1 = b \cos \beta, \qquad C_1B_1 = a \cos \alpha, \qquad B_1A_1 = c \cos \gamma.$$

In the same way the angles are found; first it is shown that

$$\sphericalangle A_1C_1B = \sphericalangle B_1C_1A = \gamma, \qquad \sphericalangle C_1A_1B = \sphericalangle B_1A_1C = \alpha,$$

$$\sphericalangle A_1B_1C = \sphericalangle C_1B_1B = \beta$$

and then it follows that

$$\sphericalangle A_1 C_1 B_1 = 180° - 2\gamma, \qquad \sphericalangle C_1 A_1 B_1 = 180° - 2\alpha,$$

$$\sphericalangle A_1 B_1 C_1 = 180° - 2\beta.$$

Let us recall the physical interpretation of minimum-length paths as light-ray or billiard-ball trajectories in Problem 4. It is then found that the minimum-perimeter orthic triangle $A_1 B_1 C_1$ is a closed triangular orbit within ABC. This suggests investigating nearby parallel paths. So let a ray of light be projected, as in Fig. 6.5c, from C_2, parallel to $C_1 A_1$. The ray reflects at A_2, B_2, C_3, A_3, and B_3, and then returns back to its starting point C_2. To show this we use the values of various angles in the figure, to prove by the sine law that

$$A_1 A_2 = C_1 C_2 \frac{\sin \alpha}{\sin \gamma}.$$

In the same way

$$B_1 B_2 = A_1 A_2 \frac{\sin \beta}{\sin \alpha}, \quad C_3 C_1 = B_1 B_2 \frac{\sin \gamma}{\sin \beta}.$$

Therefore $C_1 C_2 = C_3 C_1$ and so the path of the ray is again a closed orbit reflecting twice from each side of ABC. All such double orbits are of the same length: Since each dotted part of the orthic triangle lies halfway between two parallel parts of the double orbit, its length is the arithmetic mean of theirs. Hence the total length of the double orbit is twice the perimeter of the orthic triangle.

Problem 6. Given a triangle T with vertices A, B, C, find in the plane of T the point X that minimizes the sum $AX + BX + CX$ of vertex distances.

This problem was proposed and solved by the Italian mathematician Battista Cavalieri (1598–1647); then it was proposed by the French mathematician Pierre Fermat (1601–1665) and solved again by the Italian scientist Evarista Torricelli (1608–1647), who was an assistant to Galileo. A certain generalization from 3 to n points is known as the Steiner problem, after the German mathematician Jacob Steiner (1796–1863). A related geometrical result is known as Napoleon's theorem, after the emperor of the French.

The problem has obvious practical application whenever a communication network of roads, canals, pipes, or cables, joining together three given points, is to be shortest. The Steiner generalization results if n

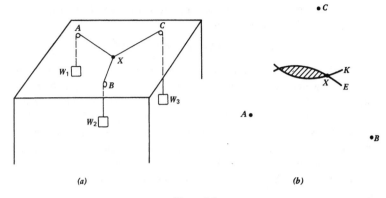

Figure 6.6

points are to be connected by the shortest network. The following mechanical analog solves the problem empirically. On the horizontal table of Fig. 6.6a there are three little pulleys at the given points A, B, C; threads tied together at X are led over them. The threads go through holes in the table and have *equal* weights W_1, W_2, W_3 tied to them. Thus X is free to move on the table, subject to the constraints of the threads and weights. In the position of equilibrium the potential energy of the system is least. Referring it to the level of the table and remembering the equality of the weights, we express the condition of minimum potential energy as $AW_1 + BW_2 + CW_3$ is maximum. But the threads do not stretch and their *total* length is fixed; hence in the equilibrium the length $AX + BX + CX$ is minimum. Therefore the position in which X comes to rest is the solution of the problem.

For the geometrical solution it is observed first that the minimizing point X cannot lie outside the triangle T. For there is then a line L that separates strictly X from A, B, and C, and if X is moved toward L all three distances AX, BX, CX are decreased. Suppose next that X lies inside T. As in Fig. 6.6b, let E be the ellipse through X with A and B as foci, and let K be the circle through X centered at C. If the ellipse and the circle intersected, as they do in Fig. 6.6b, then by moving X into the region common to E and K we save both on $AX + BX$ and on CX. Therefore E is externally tangent to K at X. Applying the theorem on the ellipse from Problem 4, we find that

$$\sphericalangle AXC = \sphericalangle BXC.$$

But the same construction may be repeated with A and C, or with B and C, as the foci of the ellipse; hence

$$\sphericalangle AXC = \sphericalangle BXC = \sphericalangle CXA = 120°.$$

This defines X as the Steiner point of the triangle $T = ABC$: the point inside T at which the sides subtend the angles 120°. Suppose, however, that T is obtuse and with an angle that is $\geq 120°$. The theorem of Thales, about all angles on a chord in a circle being equal, applies here. It follows that the longest side of T, opposite the obtuse angle, does not subtend 120° angle at *any* point inside T. Where is the minimum X then? Since it cannot lie outside T or inside T, it lies on T, and it is not hard to show that $AX + BX + CX$ is now minimized when X is the vertex of the obtuse angle.

We meet here a case of the so-called boundary extremum. A very simple case occurs in finding the extrema of the function $f(x) = x$ given for $1 \leq x \leq 2$. There is a minimum at $x = 1$ and a maximum at $x = 2$; the derivative f' is 1 everywhere and so the simple approach of equating f' to 0 runs into difficulties. Our example shows that such boundary extrema turn up in practical and uncontrived examples.

Suppose now that all angles of T are $<120°$. We show that there is then a unique Steiner point S inside T, and we give a construction for it. Let $T = ABC$ be as in Fig. 6.7a and let X be a point inside T. Rotate the triangle AXC about A as pivot by 60° into the position AYB_1. Then AXY and ACB_1 are equilateral triangles. Therefore $AX + BX + CX = BX + XY + YB_1$, and if this is to be a minimum then the points B, X, Y, B_1 are collinear. But we can repeat the same procedure at each vertex, A or B or C. It follows that the minimizing, or Steiner, point is unique: It is the point $P(60°)$ defined at the end of Chapter 2. Further, we can apply a consequence of Ptolemy's theorem, proved in Chapter 1, and we show that in Fig. 6.7b we have

$$AA_1 = BB_1 = CC_1 = AS + BS + CS.$$

For future purposes we remark that Problem 6 generalizes from 3 to n points in three completely different ways. Given n points $A_1, \ldots, A_n$ in the plane, we may ask for

1.   The point X minimizing the sum of distances $\sum_{i=1}^{n} A_i X$.
2.   The shortest connecting network that joins all the points and consists of some of the $n(n - 1)/2$ straight links $A_i A_j$.
3.   The shortest connecting network that joins all the points.

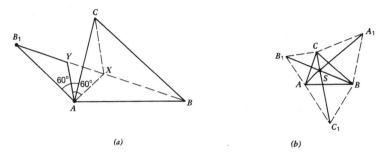

(a)   (b)

Figure 6.7

In the last formulation, in addition to the n points $A_1, \ldots, A_n$ we may introduce other vertices if necessary. The example of Problem 6 itself shows that introducing other vertices may decrease the total distance; the additional vertex was there the Steiner point S.

The next problem is different from the previous ones; in its formulation it does not even appear to be an extremum problem. The concern here is with ovals, that is, with smooth closed convex curves. As the definition of smoothness, we assume that at each point P of the curve there is a unique tangent and, therefore, a unique normal as well. If the inward-drawn normal N is considered, as in Fig. 6.8*a*, at a point P of the oval, then this normal N cuts the curve again at another point Q. It might happen that N is normal to the curve at Q as well as at P; it is then called a double normal.

Problem 7. Show that every oval has at least two double normals.

The example of an ellipse shows that there may be just two double normals and no others. As the inclusion of this problem in the present section suggests, there is a connection with extrema. To get a clue to such a connection we might reformulate our problem from a very practical point of view. Suppose that the oval C is to be clamped in an ordinary workshop vise with parallel jaws as in Fig. 6.8*b*. For instance, C might be the cross section of a length of cylindrical pipe that is being sawn. If the situation is as in Fig. 6.8*b*, then it is clear that there is a torque acting on C and so C cannot be clamped in an equilibrium position. Since the reactions F_1 and F_2 of the jaws are at right angles to the clamping planes, and the jaws are parallel, we can see the connection with Problem 7: An equilibrium position is possible if and only if the segment joining the points of contact of C with the vise is a double normal. The problem has

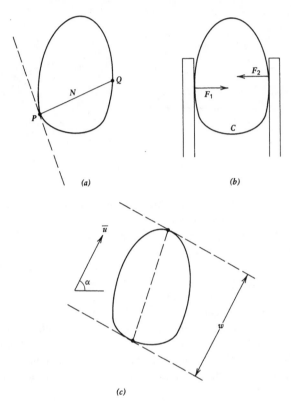

Figure 6.8

been reformulated to the following: Show that every oval can be held by a parallel-jaw vise in two different positions.

This suggests the following geometrical approach. For every direction $\bar{u}$, as given by the angle α in Fig. 6.8c, produce the two tangent lines to C, perpendicular to $\bar{u}$. Let w be the distance between them and let the segment joining the points of tangency be called the contact chord. Then w depends on $\bar{u}$ or, what is the same thing, on α, and we call it the width of C in the direction $\bar{u}$. Since w is a continuous function defined for every angle α, by a standard theorem of calculus it attains a maximum and a minimum. We verify now that the contact chords for the maximum and for the minimum widths are double normals; otherwise a small perturbation of one of the tangency points contradicts the extremum of the width.

Problem 7 may also be formulated in three dimensions, for ovoids. Here an ovoid is a smooth closed convex surface, with the same definition of smoothness as before: at each point of the ovoid there is a unique tangent plane. Therefore there is also a unique normal and so double normals are defined as before. The problem now is as follows: Show that every ovoid has at least *three* double normals. The "workshop" formulation is the following: Show that every ovoid lump can be held by a parallel-jaw vise in at least three different positions. Again the example of an ellipsoid E

$$\frac{x^2}{a^2} + \frac{y^2}{b^2} + \frac{z^2}{c^2} = 1$$

with $a > b > c > 0$ shows that there may be exactly three double normals. For E these three double normals are

The major axis: $-a \le x \le a$,
The middle axis: $-b \le y \le b$,
The minor axis: $-c \le z \le c$.

The previous procedure for ovals extends to ovoids. For an ovoid V and for a direction vector $\bar{u}$ we define the contact chord and the width $w = f(\bar{u})$ as before, replacing tangent lines by tangent planes. The size of the direction vector $\bar{u}$ is obviously irrelevant and it may be taken as 1. This defines $w = f(\bar{u})$ as a continuous function of the point $\bar{u}$ on the unit sphere. Hence f assumes a maximum and a minimum, and the corresponding contact chords give us two double normals. What about the third one?

The situation for the ellipsoid E above may be examined as a guide. Here the major axis gives the maximum width and the minor axis the minimum width. The direction of the middle axis is the Y-axis and if this direction is perturbed *in the XY-plane* to either side of the Y-axis, then the width increases. However, if the direction of the middle axis is perturbed *in the YZ-plane*, then the width decreases. Therefore the middle axis corresponds to a type of critical point of the width function known as the saddle point.

The width function $f(\bar{u})$ has the obvious property of being the same in two opposite directions: $f(\bar{u}) = f(-\bar{u})$. Although we shall not do it, it can be shown that a continuous function on the sphere with this property must have not only a maximum and a minimum but also a saddle point. Further, a double normal results not only for an extremum of $f(\bar{u})$ but also for any critical point of it. Hence there are always at least three double normals.

We return now to the original two-dimensional version of Problem 7. In proving the existence of two double normals for any oval C we have introduced the width function $f(\bar{u})$ for C. Then we argued that f attains a maximum and a minimum, giving us two double normals. What if the values of width at the maximum and at the minimum are equal? It follows then that the width is the same in every direction; consequently, every normal is a double normal. One such example is simple: when C is a circle. However, it may be surprising that there are many examples of such ovals of constant width.

Let $n \geq 3$ be an odd number—for instance, 5. We start by producing an n-gram, in our case a pentagram, as in Fig. 6.9a, in which all segments are of the same length, say 1. For this purpose we begin with three unit segments AB, BC, CD as in Fig. 6.9b, in which the angles α and β may be arbitrary, subject only to the conditions that AB crosses CD and the perpendicular bisector of AD crosses BC. On this perpendicular bisector we locate the point E so that $AE = DE = 1$, and we complete our "isosceles" pentagram. Next, with each vertex as center we describe the arc of unit circle joining the opposite two vertices, obtaining a curvilinear pentagon built of unit-circle arcs. This is shown in Fig. 6.9c; any such curvilinear polygon built of circular arcs of the same radius is called a Reuleaux polygon after a German engineer Franz Reuleaux (1829–1905), who introduced them in machine design. It is now simple to verify that every such Reuleaux polygon is a closed convex curve of which the width is the same in every direction. Also, we check that all normals are double. However, there are some difficulties because our Reuleaux polygons still have corners. One gets rid of them by the simple device of rounding shown in Fig. 6.9d: Roll a circle on the outside of the Reuleaux polygon and adjoin to that polygon the area swept out by this rolling circle. The result is an oval of constant width, with all normals double.

There are some mechanical applications of such ovals of constant width. For instance, the top rod in some fire hydrants is a rounded Reuleaux pentagon. The reason for this is to prevent vandals from turning the hydrants on, the idea being that ordinary wrenches or pliers will slide rather than grip, and the special socket used by qualified personnel is not

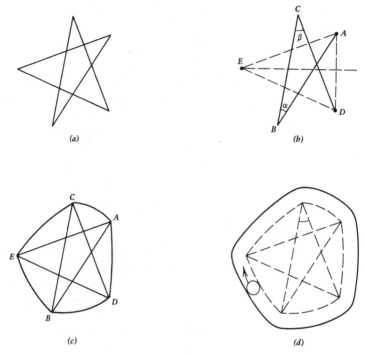

Figure 6.9

generally available. For somewhat similar reasons the new English 50 pence coin is a Reuleaux heptagon; this makes it harder to fake. Also, if we have a number of cylindrical rollers whose cross section is an oval of constant width, then on these rollers it is possible to transport large weights in the plane; because of constant width the weight neither rises nor falls when rolled on. Finally, ovals of constant width occur as essential parts in certain internal combustion engines.

Our last problem is an extremum problem but of a different type from the previous ones, and some introduction is necessary. In loose terms the problem is how to pack possibly many unit-radius circles in the plane without overlapping. There are some obvious applications—for instance, to the optimal pattern of growing fruit trees in an orchard if it is assumed that each tree sends its roots uniformly in all directions and needs a minimal area of its own. There are also less obvious and indirect applications to physics and chemistry, especially to crystallography, and to astronomy and economics. For example, there is a very direct application to packing and transportation problems when cylindrical pipes or logs are packed and moved in crates or boxcars.

To formulate the problem more precisely we consider any fixed arrangement of infinitely many unit circles in the plane, such that no two circles overlap, though they are allowed to touch. Let K be any closed curve, such as a circle, an oval, or a square; this K will be used as a gauge to measure things with. Let $K(r)$ be the result of scaling up K in the ratio $r:1$ with respect to a fixed point in K, usually some sort of a center for K. Consider then the fraction

$$\frac{\text{area of all circles inside } K(r)}{\text{area of } K(r)}$$

and let r grow indefinitely large. If the preceding fraction approaches a limit, then this limit is called the packing fraction for our arrangement of circles.

In intuitive terms the packing fraction tells us what fraction of the area of the plane is covered by the circles. The packing fraction obviously depends on the arrangement of circles we are working with, but it might also depend on the gauge K. Although we shall not prove it, within very wide limits any sufficiently regular gauge K gives the same limit; in other words, the packing fraction does not depend on K. Given two different arrangements of circles, we say that the first one is denser than the second one if the first packing fraction exceeds the second.

We work here with infinite collections of circles and so some sort of limit procedure is necessary to define packing fractions. Calculating them may be in general a matter of some difficulty. However, there is no need for any limits, and hence no calculating difficulty, in dealing with a regularly repeating arrangement of circles. One first determines a basic region B, that is, a region from which the whole infinite pattern arises by repetition. Then we divide the sum of areas of all circles, *or parts of circles*, inside B by the area of B. This quotient is exactly the packing fraction. For instance, with the regular square array of circles of Fig. 6.10*a* we may use the square shown beside it (on a larger scale) as a basic region, to get the packing fraction

$$\frac{1^2 \cdot \pi}{2^2} = 0.7854.$$

For the regular triangular array of circles of Fig. 6.10*b* either one of the two basic regions shown there may be used. By a simple calculation either

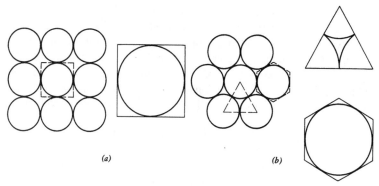

Figure 6.10

with the equilateral triangle or with the regular hexagon we get here the packing fraction

$$\frac{\pi}{\sqrt{12}} = 0.9069.$$

Problem 8. Show that the regular triangular array of circles of Fig. 6.10b has the largest possible packing fraction, $\pi/\sqrt{12}$.

The first step is to introduce a subdivision of the whole plane into certain nonoverlapping convex polygons. From analogy to tile floors such a subdivision will be called a tiling of the plane. Let $P_1, P_2, \ldots$ be any discrete collection of points in the plane, finite or infinite. _Discrete_ here means that the points do not pile up; in precise terms, no two points are closer than a certain minimum distance. With each point P_i associate the region V_i consisting of all points closer to P_i than to any other point P_j. For technical reasons _closer to_ is changed to _no further from_; this has the advantage of giving closed regions V_i: the boundary of V_i is now included in V_i.

The region V_i is called the ith Voronoi region, or the Voronoi region of the point P_i, after a Russian mathematician G. Voronoi (1866–1908). Sometimes it is also called the Dirichlet cell of P_i, after a German mathematician P. L. Dirichlet (1805–1859). In symbols and with the ordinary terminology for sets

$$V_i = \{P: PP_i \le PP_j \quad \text{for every} \quad j\}.$$

Let us study some examples. For only two points P_1 and P_2 the Voronoi regions V_1 and V_2 are the two half-planes with a common boundary that is the perpendicular bisector of the straight segment P_1P_2. With three points P_1, P_2, P_3 there are two cases illustrated in Fig. 6.11a and b. If the three points are not collinear, then V_1, V_2, V_3 are three angular regions of Fig. 6.11a, meeting at a point A. Since this A is equidistant from P_1, P_2, P_3, it is the circumcenter of the triangle $P_1P_2P_3$. If P_1, P_2, P_3 are collinear, then the Voronoi regions V_1, V_2, V_3 are an infinite strip and the two adjoining half-planes as in Fig. 6.11b. An example with four points

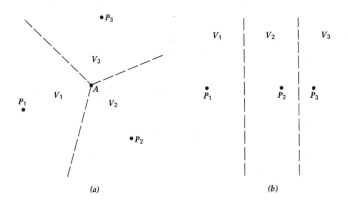

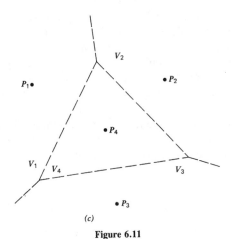

Figure 6.11

is shown in Fig. 6.11c. As these examples suggest, each point of the plane belongs to at least one V_i, and each V_i is a convex polygonal region, bounded by pieces of the perpendicular bisectors P_1P_i, P_2P_i, ...

Let $H(i, j)$ denote the half-plane containing P_i and bounded by the perpendicular bisector of P_iP_j. Then

$$V_i = \bigcap_{\substack{\text{all } j \\ j \neq i}} H(i, j) \tag{1}$$

This merely rephrases in symbols the previous description of the Voronoi region V_i in words. The half-plane $H(i, j)$ is the locus of all points no further from P_i than from P_j. The right-hand side of (1) is the set of points common to all such half-planes, that is, the set of points no further from P_i than from *any* P_j. Hence it is V_i. By (1) each V_i is an intersection of half-planes; it is therefore a convex polygonal region.

We start now on the proof of the proposition contained in Problem 8: The regular triangular array of nonoverlapping unit circles is the densest possible. That is, given *any* array of nonoverlapping unit circles C_1, C_2, . . . , its packing fraction is at most $\pi/\sqrt{12}$. Let $P_1, P_2, \ldots$ be the centers of our circles; no two are closer than 2 and so the collection P_1, P_2, . . . is discrete. Let $V_1, V_2, \ldots$ be the tiling of the whole plane for the points $P_1, P_2, \ldots$. Since the collection $P_1, P_2, \ldots$ is discrete and since we want to produce a possibly dense packing of the circles $C_1, C_2, \ldots$, it follows that each region V_i is a convex polygon (see problems 9 and 11 in the Exercises). It has then a well-defined area $\mathrm{Ar}(V_i)$. Also, it is obvious that each Voronoi region V_i contains its circle C_i. Hence

$$\text{packing fraction} = \lim_{n \to \infty} \frac{n\pi}{\sum_{i=1}^{n} \mathrm{Ar}(V_i)} \tag{2}$$

Our immediate aim is to obtain the basic estimate

$$\text{for every } i \quad \mathrm{Ar}(V_i) \geq \sqrt{12} \tag{3}$$

Once this is proved then by (2) the packing fraction is necessarily $\leq \pi/\sqrt{12}$. But the regular triangular array achieves the packing fraction $\pi/\sqrt{12}$; hence this fraction is the best possible.

We go on now to prove (3). Consider the Voronoi region V_i belonging to the point P_i. It has an interior; this consists of all points *strictly* closer to P_i then to any P_j, $j \neq i$. It also has a boundary consisting of straight segments. These segments contain points equidistant from P_i and from some other point P_j. Finally, consider a vertex A of the region V_i. As in Fig. 6.11a, this vertex A is equidistant from some three points: P_i and two others, say P_j and P_k. It is therefore the circumcenter of the triangle $P_iP_jP_k$. Let the sides of this triangle be a, b, c and the angles opposite them α, β, γ; also let R be the circumradius. Then by the sine law

$$\frac{a}{\sin \alpha} = \frac{b}{\sin \beta} = \frac{c}{\sin \gamma} = 2R.$$

But a, b, c are each ≥ 2 since the circles C_i, C_j, C_k do not overlap. Also,

$$\min(\alpha, \beta, \gamma) \leq 60°$$

since $\alpha + \beta + \gamma = 180°$. Since $\sin 60° = \sqrt{3}/2$ and $\sin x$ increases with x when $0° \leq x < 90°$, it follows that $R \geq 2/\sqrt{3}$. This means that any vertex of V_i cannot be closer to P_i than $2/\sqrt{3}$.

We put now together all the necessary information about V_i. Let D_i be the circle of radius $2/\sqrt{3}$ about P_i as center; then C_i and D_i are concentric, and

1. V_i is a convex polygon.
2. V_i contains C_i.
3. The vertices of V_i lie on or outside D_i.

These three pieces of information will be enough to prove the required estimate (3). The idea of the proof is as follows. First, it is shown that without loss of generality item 3 may be replaced by

3a. The vertices of V_i lie on D_i.

Without loss of generality here means "without increase of area." This makes of V_i a convex polygon containing C_i and inscribed into D_i. To have possibly small area, such a polygon must have the fewest possible sides. Now a simple calculation shows that the regular hexagon inscribed into D_i just contains C_i and happens to have the area exactly $\sqrt{12}$.

The details of the proof may be arranged as follows. First, to replace item 3 by 3a we consider Fig. 6.12a, erasing the vertex A outside D_i and replacing it by two new vertices A_1 and A_2 on D_i. If the side A_1A_2 is so long that it cuts C_i, then new extra vertices are added on D_i, between A_1 and A_2. Certainly this will not increase the area of V_i. Next, of any two consecutive sides at least one may be taken as tangent to C_i. To show

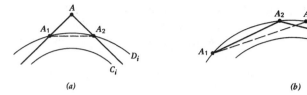

Figure 6.12

this we consider Fig. 6.12b where $A_1A_2 \leq A_2A_3$. Then A_2 is moved on D_i toward A_1. Either it can be moved all the way to A_1 itself and so the side A_1A_2 is eliminated or the process stops when A_2A_3 becomes tangent to C_i. Again, the area of V_i is never increased. Finally, any two consecutive sides may be switched around: For Fig. 6.12b the vertex A_2 is replaced by A_2' such that $A_1A_2 = A_2'A_3$ and $A_1A_2' = A_2A_3$; here the area of V_i remains the same.

By a sequence of such transpositions we can arrange V_i so that all the sides that are tangent to C_i are consecutive. Are there any other sides left? No, for by the preceding there cannot be two consecutive sides left, and one such side will not make V_i close up. Therefore we are left with a regular hexagon of area $\sqrt{12}$. Hence the original polygon V_i has area $\geq \sqrt{12}$ and we are finished.

We end with a brief mention of packings in space. By analogy with the plane there are now packings, and packing fractions, by nonoverlapping unit-radius spheres. The definition of a Voronoi region carries over without any real change. In analogy to the regular triangular packing in the plane, there is now the regular tetrahedral packing, for which the packing fraction is $\pi/\sqrt{18} = 0.7404$. Although it is almost certain that this is the best packing fraction possible in three dimensions, it may come as a surprise that this has never been proved.

EXERCISES

1. For problem 1, and the case illustrated in the middle of Fig. 6.1b, show that the intersection point is an admissible center of rotation. (*Hint*: Show that the feet of three perpendiculars to the sides of a triangle T from a point p *inside* T can be vertices of a rectangle R lying in T but that this is impossible if the point p lies *outside* T.)

2. In problem 1 attempt to show the maximality of the half-height triangle directly, without the rotation argument. (*Hint*: Show that if a thin rectangular parcel R lies over a triangular sheet of paper, then R can be wrapped up.)

3. If all three maximal rectangles of problem 1 are drawn in the triangle T, show that one quarter of T's area is covered triply, one quarter doubly, one quarter singly, and one quarter not at all.

4. Let T be a triangle with circumradius R and in-radius r, and let T_1 be a triangle inscribed into T. Let L and L_1 be the perimeters of T and of T_1. Show that $L_1 \geq rL/R$.

5. Give an example for Miquel's theorem where the three circles intersect outside the triangle and prove the theorem for that case.

6. Attempt to generalize problem 3 to four or more points. (*Hint*: Start with four circles passing through one point.)

7. Let V be any convex n-gon and P any point inside V. Find points $P_1, P_2, \ldots, P_n$ such that in the configuration of the $n + 1$ points $P, P_1, P_2, \ldots, P_n$ the Voronoi region of P is V.

8. For the regular square array of Fig. 6.10a show without *any* calculations that the packing fraction can be improved. (*Hint*: Consider every second row of circles.)

9. Let $P_1, P_2, \ldots$ be any sequence of points in the plane, Show that a necessary and sufficient condition for the existence of an unbounded Voronoi region V_i is that the plane has a half-plane free of the points.

10. Give an example of a sequence of points $P_0, P_1, P_2, \ldots$ such that the Voronoi region of P_0 is a bounded convex region but not a polygon. (*Hint*: Take P_0 to be the center of a circle C and Q_1 to be a point on C, then any point on C is given by the angle at P_0, taking Q_1 with the zero-angle; choose an infinite sequence $Q_2, Q_3, \ldots$ of points on C such that the corresponding angles increase steadily, with the limit 2π. Now produce $P_1, P_2, \ldots$ so that the Voronoi region of P_0 has $Q_1, Q_2, \ldots$ as its set of vertices.)

11. Prove that the preceding case cannot occur if the sequence $P_1, P_2, \ldots$ is discrete. That is, show that for a discrete sequence of points a bounded Voronoi region is a polygon.

12. Two points A and B lie on the same side of a straight line L. Produce the shortest network joining A and B to L, distinguishing between various cases.

13. Let the triangle ABC have sides a, b, c, area D, and angles $<120°$. Let d be the minimum value of the sum $AX + BX + CX$. Using the cosine law in Fig. 6.7a show that

$$2d^2 = a^2 + b^2 + c^2 + 4\sqrt{3}D.$$

14. Let S be the Steiner point in the preceding example and let it have distances x from A, y from B, z from C. Show that

$$x = \frac{1}{3}\left(d + \frac{b^2 + c^2 - 2a^2}{d}\right), \quad y = \frac{1}{3}\left(d + \frac{a^2 + c^2 - 2b^2}{d}\right),$$

$$z = \frac{1}{3}\left(d + \frac{a^2 + b^2 - 2c^2}{d}\right).$$

15. Let a, b, c be the sides of a triangle and D its area. Show that

$$a^2 + b^2 + c^2 - 4\sqrt{3}D \geq 0$$

and the equality holds only for the equilateral triangles. (*Hint*: minimize $b^2 + c^2$, keeping a and D fixed; conclude that $a^2 + b^2 + c^2 - 4\sqrt{3}D$ is least for equilateral triangle.)

16. Combining problems 13 and 15, show that

$$4\sqrt{3}D \leq d^2 \leq a^2 + b^2 + c^2.$$

17. Let the triangle $T = ABC$ have vertex angle at $A \geq 120°$. Show that $AX + BX + CX$ is minimum when $X = A$. (*Hint*: X lies on the periphery of T; show that there are only two candidates for X: A and its projection on the opposite side.)

18. Prove that no matter where the point X lies inside an equilateral triangle T, the sum of its distances to the three sides of T is constant.

19. Let the triangle $T = ABC$ have all angles less than $120°$. Use problem 18 to give a new proof that the sum $AX + BX + CX$ is minimum when X is the Steiner point of T. (*Hint*: Produce a suitable equilateral triangle T and apply problem 18.)

CHAPTER SEVEN

SIMPLE GEOMETRY
AND TRIGONOMETRY
ON THE SPHERE

The greatest mathematician and scientist of antiquity was Archimedes (287–212 B.C.), a Greek who lived in the Greek city-state of Syracuse in Sicily. Among his discoveries are the basic law of hydrostatics on the floating bodies, the basic law of statics on the lever, a computation of π together with a method that allows one in principle to compute π to arbitrary accuracy, a formula for the area of a parabolic segment, a proof of the formula for the area of a sphere, and many others.

His hydrostatic discovery is supposed to have been made in connection with the gold crown for the king of Syracuse. It was suspected that the goldsmith had cheated by substituting the cheaper gold-silver alloy for the gold given to him. Archimedes is reputed to have been thinking about the problem while taking a bath and the idea of the law of floating bodies, and its relevance to the matter, occurred to him then. According to the story, he ran naked into the street shouting "Eureka, eureka" (I have found it, I have found it).

His law of lever equilibrium states that the products of the force and the distance from the fulcrum are equal. Here Archimedes is reputed to have said, "Give me whereon to stand and I shall move the earth."

Archimedes proved that the area of a sphere equals the lateral area of the cylinder that wraps it around. He must have valued this discovery very highly since a cylinder wrapping a sphere was engraved on his tombstone. Some 150 years later, when Sicily belonged to Rome, Marcus Tullius Cicero was an official in Sicily and the location of Archimedes' tomb was then unknown. However, by following the tradition, Cicero was able to locate a pile of stones in a cypress grove, with a sphere and a cylinder engraved on them.

There are other examples in history of a mathematician's chief work

123

being marked on his tomb. For instance, the Dutch mathematician Ludolph van Ceulen (1540–1610), a professor at the University of Leyden in Holland, spent many years of his life calculating π to 35 decimals. This fact, together with the value of π, was stated on his tombstone in St. Peter's church in Leyden. The Swiss mathematician Jacob Bernoulli (1654–1705), impressed by various geometrical properties of the logarithmic spiral, ordered that it should be engraved on his tombstone with the Latin phrase *Eadem mutata resurgo* (However changed, I return). The Italian mathematician G. T. di Fagnano (the father of the Fagnano of Problem 5 in the last chapter) (1682–1766), made similar arrangements with respect to another curve, the lemniscate. The great German mathematician Carl Friedrich Gauss (1777–1855) made in his youth the striking discovery that a regular polygon of 17 sides can be constructed by ruler and compasses. Gauss is reputed to have requested that such a polygon be engraved on his tomb, and to have given up the idea because the stonemason had thought that the polygon would be indistinguishable from a circle.

We turn now to the Archimedean proposition that the area of a sphere S equals the area of its wrapping cylinder, and we prove a more general theorem. Let the sphere S be as in Fig. 7.1 and let C be the vertical cylinder that wraps round S and is of the same vertical extent as S. Then S and C touch in the horizontal great circle E on the sphere S, which is

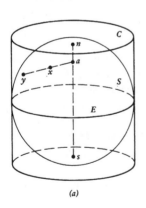

(a)

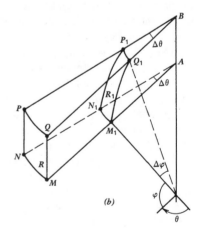

(b)

Figure 7.1

the equator of S. Let n and s be the corresponding north and south poles of S; the straight vertical segment ns is the common axis of S and of C. Let x be any point of S, other than the poles n and s. There is then a unique point a on the axis ns at the same height as x; it is the point on ns that is closest to x. When the straight horizontal segment ax is produced beyond x, its continuation will cut the cylinder C at some point y, at the same height as x and a. If x happens to lie on E, then x and y coincide.

We call y the axial projection of x. This projection A defines a correspondence between sets on S and sets on C: To any set X of points on S there corresponds its axial image $A(X)$ on C consisting of the axial projections of all points in X. The generalized theorem of Archimedes states that axial projection preserves areas: For any X

$$\text{area } X = \text{area } A(X). \tag{1}$$

It is assumed here that the sets X and $A(X)$ are sufficiently regular to have areas. That is, we suppose that they can be divided into a large number of some simple small elementary parts; we add up their areas and pass to the limit as is usual with setting up integrals. The poles n and s are anomalous: They have no unique axial projections. However, we may take their axial projections to be the whole upper and lower circular rims of the cylinder C. Since the points n and s on S, as well as the circular rims on C, have zero area, the exceptional role of n and s presents no difficulty in (1).

As a particular case of (1) X may be the whole sphere S, $A(X)$ is then the whole cylinder C, and (1) asserts then the original form of Archimedes' theorem: Area S = area C. If C is slit open along any vertical generator, it can be unrolled onto the plane as a rectangle R. Let r be the radius of S, then the height of R is $2r$ and its base is the circumference $2\pi r$ of the equator E. This unrolling is clearly area-preserving and so the area of S is $4\pi r^2$. An equivalent statement is the exact original form in which Archimedes has expressed it; the Greeks did not operate with algebraic quantities and formulas, and so Archimedes has put it as follows: The area of a sphere equals four times the area of its greatest circle.

There is in a Moscow museum an ancient Egyptian mathematical manuscript, preceding Archimedes by more than 1000 years. It is concerned with practical things such as calculation of areas and volumes, and it seems to state that the area of a hemisphere is twice the area of its cover circle. Since the ancient Egyptian geometry was rather primitive it is possible that the preceding assertion, though equivalent to the theorem of Archimedes, was really an empirical observation. The ancient Egyptians were skilled at weaving and they wove reeds as well as yarn; it may have been observed that in weaving a hemispherical reed basket twice as much material is needed for the basket as for its circular cover.

We return now to the geometry of our sphere S and cylinder C of Fig. 7.1a. By analogy to geography any semicircle on S, from the north pole n to the south pole s, is called a meridian of S. Also, any horizontal circle on S is called a latitude circle. In particular, the equator E is a latitude circle, the only one that is a great circle of the sphere S (i.e., has the same radius as it). All other circles of latitude have radii smaller than S, and so they are small circles of S.

From our definition of axial projection it follows that any meridian of S projects axially onto C as a straight vertical segment on C. A latitude circle on S projects axially onto C as a horizontal circle on the cylinder C. So it is noted that all latitude circles of S, no matter of what size, project axially onto C as circles of the same size. In fact, we get their projections just by moving E up or down on C.

We come now to the proof of (1). Let X be a region on the sphere S and $A(X)$ its axial projection on the cylinder C. The areas of X and of $A(X)$ are obtained by the standard technique of integral calculus: Each region is subdivided into a large number N of some convenient small elements, their areas are added up, and then the limit is taken as N tends to infinity. Our proof hinges on a suitable choice of those area elements on S and on C. For this purpose we use on the sphere S the meridians and the circles of latitude. Clearly, as many of either can be drawn as is needed, so that X gets subdivided into arbitrarily small parts. On the cylinder C we use the axial projection of the meridians and latitude circles. Thus any subdivision of X on S will automatically induce a corresponding subdivision of $A(X)$ on C.

Consider now a pair of such elements of area, one on S and one on C, shown in Fig. 7.1b. On the sphere S the area element is a curvilinear rectangle $R_1 = M_1 N_1 P_1 Q_1$ bounded by two small congruent meridian arcs $\widehat{M_1 Q_1}$ and $\widehat{N_1 P_1}$ and by two small arcs of two different latitude circles, $\widehat{N_1 M_1}$ and $\widehat{P_1 Q_1}$. Let ϕ be the latitude and θ the longitude of the point M_1; then the latitudes and longitudes of N_1, P_1, Q_1 are

$$\phi, \ \ \theta + \Delta\theta; \quad \phi + \Delta\phi, \ \ \theta + \Delta\theta; \quad \phi + \Delta\phi, \ \ \theta.$$

On the cylinder C the coresponding area element is a curvilinear rectangle R bounded by two small straight vertical segments MQ and NP that are congruent and by two small circular arcs $\widehat{NM}$ and $\widehat{PQ}$, also congruent. Its area is given *exactly* by

$$\text{area } R = MQ \cdot \text{length } \widehat{NM}$$

so that

$$\text{area } R = r^2\, \Delta\theta[\sin(\phi + \Delta\phi) - \sin\phi] \tag{2}$$

where r is the radius of S.

The area of R_1 is given *approximately* as

$$\text{area } R_1 = \text{length } \widehat{N_1 M_1} \cdot \text{length } \widehat{M_1 Q_1} + \epsilon$$

or, in terms of the latitude and longitude angles, as

$$\text{area } R_1 = r^2 \cos\phi\, \Delta\theta\, \Delta\phi + \epsilon. \tag{3}$$

Here ϵ is an error term that is negligible, *relative to* $\Delta\theta\, \Delta\phi$, in the limit when $\Delta\theta$ and $\Delta\phi$ approach 0. By the mean value therorem (2) may be written as

$$\text{area } R = r^2 \cos\phi\, \Delta\theta\, \Delta\phi + \epsilon_1 \tag{4}$$

where ϵ_1 is an error term, negligible in the same sense as ϵ. Now (3) and (4) assert that the area elements on S and on C are the same. Hence (1) is proved: X and $A(X)$ have equal areas since these two areas are obtained by integrating the two equal area elements.

To sum it up briefly, and somewhat inaccurately, the elementary rectangles R and R_1 have the same area because on passing from R to R_1 the base gets compressed in the same ratio in which the height gets stretched. This ratio is the cosine of the latitude angle ϕ.

It was shown already as a consequence of (1) that area $S = 4\pi r^2$. Somewhat more generally, (1) enables us to find the areas of certain simple subsets of S. For instance, let $L(\alpha)$ be a spherical lune of angle α; this is the part of the sphere S contained between two meridians making the angle α. Here α may be taken as the angle at which the meridians cut at either pole, n or s, or, equivalently, as the angle subtended at the center of S by that arc of the equator E, which lies inside the lune. A simple application of (1) shows that

$$\text{area } L(\alpha) = 2\alpha r^2. \tag{5}$$

Again, suppose that $0 \le \alpha \le 90°$ and let $P(\alpha)$ be the spherical cap of angular radius α; this is that part of the sphere S that lies north of the latitude $90° - \alpha$. Here an application of (1) shows that

$$\text{area } P(\alpha) = 2\pi r^2(1 - \cos \alpha). \tag{6}$$

As a less immediate and less obvious application of (1) we consider a problem in probability. This problem is known as the continuous random walk and is of certain theoretical and practical importance; it is as follows. Let $\bar{u}_1, \ldots, \bar{u}_n$ be n vectors in the three-dimensional Euclidean space. The vectors are taken independently of each other; every one is of unit length and its direction is uniformly at random. The problem is the following: What is the probability distribution of the resulting vector sum

$$\bar{v} = \bar{u}_1 + \cdots + \bar{u}_n ?$$

By the complete spherical symmetry the *direction* of $\bar{v}$ is itself uniformly at random, and the thing of interest is the probability distribution of the *magnitude* $| \bar{v} |$.

It is necessary to examine first in some detail the assumption that a vector $\bar{u}$ of fixed magnitude has uniformly random direction: We want to express the intuitive notion that $\bar{u}$ is as likely to point in one direction as in any other. Let $\bar{u}$ start from the origin O, then its tip lies on the sphere S whose radius is the magnitude $| \bar{u} |$. The uniformly random direction means that the tip of $\bar{u}$ is just as likely to lie in X as in Y, whenever X and Y are two subsets of the sphere of the same area. Equivalently, the probability that $\bar{u}$ terminates in X is the ratio area X/area S.

Already the appearance of the quantity Area X suggests a connection with the generalized Archimedean theorem (1). To develop such a connection we argue as follows. The direction of the vector $\bar{u}$ is uniformly at random and so its tip is equidistributed over the sphere S of radius $| \bar{u} |$. Let D be any diameter of S; we claim that the projection of $\bar{u}$ onto D is equidistributed on the diameter D. To show this, we must prove that if I_1 and I_2 are any two intervals on D, of the same length, then the projection of u onto D is as likely to lie in I_1 as in I_2. However, this is a simple consequence of (1) and of the very definition of axial projection. All that is needed is to observe that by (1) the area of the spherical zone on S, bounded by two circles of latitude, depends only on the distance between the planes containing those circles and *not* on the position of the circles on S.

This supplies the crucial reduction of our random-walk problem from *three* dimensions to *one*: The probability distribution of $| \bar{v} |$ is the same as the probability distribution of $|x|$ where

$$x = x_1 + \cdots + x_n;$$

here the variables $x_1, \ldots, x_n$ are independent, and each is a scalar equidistributed on the interval $[-1, 1]$. Now this is a standard question in probability theory and is (fairly) simply answered.

In the next problem considered we keep the same sphere S of radius r and we ask, "What is the area of a spherical triangle T on S, with given angles α, β, γ?" Such a spherical triangle has three vertices A, B, C, on S and its sides are three great-circle arcs $\widehat{AB}$, $\widehat{AC}$, $\widehat{BC}$; α, β, γ are the angles at which the sides meet, as shown in Fig. 7.2a. If the great-circle arcs $\widehat{AC}$ and $\widehat{AB}$ are continued beyond C and B, they meet again at the point A' antipodal to A. The extended arcs are meridians and they enclose the lune $L(\alpha)$. The set of all points antipodal to points in $L(\alpha)$ forms the antipodal lune $L'(\alpha)$. Consider now the six lunes

$$L(\alpha), \quad L'(\alpha), \quad L(\beta), \quad L'(\beta), \quad L(\gamma), \quad L'(\gamma). \qquad (7)$$

By (5) the sum of their areas is

$$4r^2(\alpha + \beta + \gamma). \qquad (8)$$

The essence of our method is that the same quantity—the sum of areas of the six lunes—can also be computed in another way. It is observed that the six lunes together cover the whole of S; more precisely, every point on S is covered exactly once except the points of T and of the

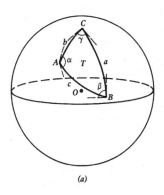

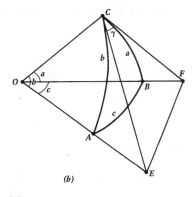

(a)   (b)

Figure 7.2

antipodal triangle T'; these are covered triply. Hence the sum of the areas of our six lunes in (7) is

$$4\pi r^2 + 4 \text{ area } T. \tag{9}$$

Here we have used repeatedly the obvious fact that a set on S and its antipodal set have equal areas. Now, equating (8) and (9), we get

$$\text{area } T = r^2(\alpha + \beta + \gamma - \pi). \tag{10}$$

In words, the area of T is r^2 times the spherical excess of the sum of the angles of T over π. Since the area of any proper spherical triangle T, that is, a triangle on S that does not collapse to an arc, is necessarily positive, the following important consequence results from (10):

In a spherical triangle the three angles always add up to more than 180°.

There is another way of expressing (10), by means of the so-called steradian content. To motivate this concept we recall that an angle in the plane is conveniently measured by its radian content: A circle is described with the vertex of the angle as center and the arc length contained in the angle is divided by the radius r of the circle. The radius r is arbitrary; the *same* radian content is obtained for every $r > 0$. To generalize this to three dimensions, it is first necessary to replace the angle by a solid angle. This is simple since a solid angle is just a solid cone. Now the steradian content is defined as before. A sphere S of arbitrary radius r is centered at the vertex of the cone; let A be the spherical figure cut out of S by the cone, then the steradian content is the area of A divided by r^2. Of course, any cone can be taken here, not just the "ordinary" circular cone. For instance, with reference to Fig. 7.2a we could take as our cone the solid trihedral angle subtended by T at the center O. This consists of all rays emerging from O and crossing T. Now (10) tells us that the steradian content of the solid trihedral angle is $\alpha + \beta + \gamma - \pi$.

The rest of this section is devoted to developing the fundamentals of spherical trigonometry. This subject deals with spherical triangles just as the ordinary, i.e., plane trigonometry, deals with plane triangles. In fact, the word *trigonometry* means, according to its etymology, "measurement of triangles." There is no need to justify the study of spherical trigonometry in view of its obvious applications to geography and astronomy.

Let T be a spherical triangle—for instance, that of Fig. 7.2a. Its three vertex angles or briefly angles α, β, γ have already been defined. We define next its *sides a, b, c* as the *angles* at which the arcs $\overset{\frown}{BC}, \overset{\frown}{AC}, \overset{\frown}{AB}$ appear at the center O of the sphere S on which T lies. This might appear to be strange: Should one not define the sides rather as the *lengths* of those three arcs? If these three lengths are u, v, w, the three angles, in radians, are simply $a = u/r$, $b = v/r$, $c = w/r$, where r is the radius of our sphere. Thus there is really no difference whether the sides are defined as angles or as lengths. The usual tradition of taking angles has the advantage of giving formulas that are free of the radius r of the fixed sphere; there is less writing to do.

A spherical triangle T has now six *elements*: the three angles α, β, γ and the three sides a, b, c. The usual convention is observed: a is opposite α, b opposite β, c opposite γ. It will be always assumed that

T lies completely inside a hemisphere of S.

An equivalent assumption is that

All six elements of T are less than 180°.

Our object is to develop sufficient apparatus of spherical trigonometry to be able to solve any spherical triangle T. The following is meant by solving T. A spherical triangle T has six elements but it is given by any three of them (except for certain special cases that will be stated); to solve T means to find the other three elements in terms of the three given ones. A fundamental difference between the plane and the spherical trigonometry must be emphasized here. In the plane a triangle is *not* given when its three angles are known. This is so because in the plane there exist *similar* triangles, any two of which have the same angles. In spherical trigonometry we work with spherical triangles of a fixed sphere S and here the whole concept of similarity disappears. In particular, a triangle on S is given uniquely by its three angles.

However, even though similarity is lost on the sphere, it will be seen that something is gained. This will be the duality, or more precisely, the polar duality, a concept for spherical triangles that does not apply to plane triangles.

One of the principal results in plane trigonometry is the cosine law: If a and b are two sides of a plane triangle and γ is the angle between them, then the third side c is given by

$$c^2 = a^2 + b^2 - 2ab \cos \gamma \tag{11}$$

The special case $\gamma = \pi/2$ is the Pythagorean theorem. We prove now the *spherical* cosine law. Let a and b be two sides of a spherical triangle and γ the angle between them; suppose first that $a < \pi/2$ and $b < \pi/2$. As is shown in Fig. 7.2b, we draw the tangents to $\overset{\frown}{AC}$ and to $\overset{\frown}{BC}$ at C; let these tangents cut the plane OAB in E and in F. Here O is the center of our basic sphere, and r is its radius. Since OCE and OCF are right-angled triangles, and $OC = r$, we have

$$OE = r \sec b, \quad OF = r \sec a, \quad CE = r \tan b, \quad CF = r \tan a. \quad (12)$$

EF is next computed twice by applying the *plane* cosine law, once to $\triangle OEF$ and once to $\triangle CEF$; we get

$$EF^2 = r^2 \sec^2 b + r^2 \sec^2 a - 2r^2 \sec a \sec b \cos c,$$

$$EF^2 = r^2 \tan^2 b + r^2 \tan^2 a - 2r^2 \tan a \tan b \cos \gamma.$$

Equating these, recalling that $\sec^2 x = 1 + \tan^2 x$, and simplifying, we get

$$\cos c = \cos a \cos b + \sin a \sin b \cos \gamma, \quad (13)$$

which is the desired spherical cosine law. The two analogous expressions are

$$\cos b = \cos a \cos c + \sin a \sin c \cos \beta. \quad (14)$$

$$\cos a = \cos b \cos c + \sin b \sin c \cos \alpha. \quad (15)$$

This proves the spherical cosine law but subject still to the restrictions $a < \pi/2$, $b < \pi/2$. If these do not hold, then we cannot use our argument. However, (13) holds without any restriction on a and b; we show it by using the lune diagrams. Suppose first that $a > \pi/2$ while $b < \pi/2$. Let ABC be our spherical triangle and extend the arcs $\overset{\frown}{BC}$ and $\overset{\frown}{BA}$ until they meet at the point B' antipodal to B, as is shown in the lune diagram of Fig. 7.3a. Then $AB'C$ is a spherical triangle colunar to ABC. Its sides meeting at C are $\pi - a$ and b, and *both* are $<\pi/2$. Therefore the preceding proof applies; the angle between those two sides is $\pi - \gamma$. Now, applying the spherical cosine law to $AB'C$, we show that (13) still holds:

$$\cos(\pi - c) = \cos(\pi - a)\cos b + \sin(\pi - a)\sin b \cos(\pi - \gamma),$$

which after simplification gives us (13) again. Similarly, if both $a > \pi/2$ and $b > \pi/2$, we apply the spherical cosine law to the colunar triangle

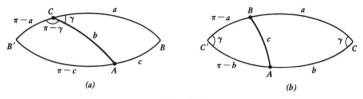

Figure 7.3

ABC' of Fig. 7.3*b*. The cases when one or both sides *a*, *b* are $= \pi/2$ are left as exercises.

It will be noticed that the spherical cosine law (13) has been deduced from the plane cosine law (11). The reverse can also be done; to do this we start with (13) but we write a/r for a, b/r for b and c/r for c, operating now with *lengths* as sides. This gives us

$$\cos\frac{c}{r} = \cos\frac{a}{r}\cos\frac{b}{r} + \sin\frac{a}{r}\sin\frac{b}{r}\cos\gamma. \tag{16}$$

To obtain the plane case, we assume that $r \to \infty$ and in the limit the spherical triangle becomes the plane triangle. Recall that for x small

$$\cos x \doteq 1 - \frac{x^2}{2} + \ldots, \qquad \sin x = x - \cdots$$

and apply these expansions to (16) getting

$$1 - \frac{c^2}{2r^2} + \cdots = \left(1 - \frac{a^2}{2r^2} + \cdots\right)\left(1 - \frac{b^2}{2r^2} + \cdots\right)$$
$$+ \left(\frac{a}{r} - \cdots\right)\left(\frac{b}{r} - \cdots\right)\cos\gamma,$$

multiplying out and simplifying we get in the limit when $r \to \infty$

$$c^2 = a^2 + b^2 - 2bc\cos\gamma,$$

which is the plane cosine law (11).

We next obtain the spherical law of sines. Starting with (13) in the form

$$\cos\gamma = \frac{\cos c - \cos a \cos b}{\sin a \sin b},$$

we compute

$$\sin^2 \gamma = 1 - \cos^2 \gamma = \frac{\sin^2 a \sin^2 b - (\cos c - \cos a \cos b)^2}{\sin^2 a \sin^2 b}$$

$$= \frac{(1 - \cos^2 a)(1 - \cos^2 b) - (\cos c - \cos a \cos b)^2}{\sin^2 a \sin^2 b}$$

$$= \frac{1 - \cos^2 a - \cos^2 b - \cos^2 c + 2 \cos a \cos b \cos c}{\sin^2 a \sin^2 b}.$$

Therefore, taking square roots

$$\frac{\sin \gamma}{\sin c} = \frac{[1 - \cos^2 a - \cos^2 b - \cos^2 c + 2 \cos a \cos b \cos c]^{1/2}}{\sin a \sin b \sin c}, \quad (17)$$

where only the positive root is taken (since $\sin x > 0$ for $0 < x < \pi$). However, the R.H.S. of (17) is a *symmetric* function of a, b, c: *therefore*

$$\frac{\sin \alpha}{\sin a} = \frac{\sin \beta}{\sin b} = \frac{\sin \gamma}{\sin c}$$

$$= \frac{[1 - \cos^2 a - \cos^2 b - \cos^2 c + 2 \cos a \cos b \cos c]^{1/2}}{\sin a \sin b \sin c}, \quad (18)$$

which is the spherical law of sines.

We take up next the previously mentioned polar duality. Let $T = ABC$ be a spherical triangle on S. Its three arcs $\widehat{AB}$, $\widehat{AC}$, $\widehat{BC}$ are great-circle arcs on S and therefore they lie on three equators of S. Let A_1 be the pole of the equator BC, B_1 the pole of the equator AC, C_1 the pole of the equator AB. A certain care is needed here: Each equator has *two* poles, which one are we to take? Recall the convention that the triangle ABC is contained in some hemisphere H of S; now for each pair of poles we take that one that lies also in H. With this the poles A_1, B_1, C_1 are unambiguously given, and we call the new triangle $T_1 = A_1B_1C_1$ the polar dual of T, or briefly its polar triangle. It turns out that the relation is reciprocal (which justifies the term *duality*): the triangle polar to T_1 is T itself again. In other words, the polar of the polar of a triangle T is T itself.

To show this consider T and T_1 as drawn in the diagram of Fig. 7.4a. A_1 is by definition the pole of $\widehat{BC}$, it may lie inside T (as in the figure) outside T, or even on T; similarly for B_1 and C_1. Since B_1 is the pole of $\widehat{AC}$ the great-circle arc $\widehat{B_1A}$ is a quadrant. Also, C_1 is the pole of $\widehat{AB}$ and therefore the great-circle arc $\widehat{C_1A}$ is also a quadrant. Now $\widehat{B_1A}$ and $\widehat{C_1A}$

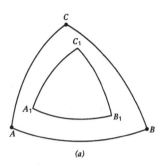
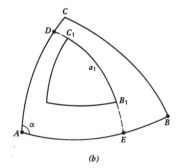

      (a)   (b)

Figure 7.4

are both quadrants, which implies that

$$A \quad \text{is the pole of} \quad \widehat{B_1C_1}.$$

Analogous reasoning shows that B is the pole of $\widehat{A_1C_1}$ and C is the pole of $\widehat{A_1B_1}$. Hence the polar triangle of T_1 is ABC or T itself, q.e.d.

The new, polar, spherical triangle T has its six elements: the three sides a_1, b_1, c_1 and the three angles α_1, β_1, γ_1. It will be shown that the six new elements are simply related to the six old ones. For this purpose consider the diagram of Fig. 7.4b showing $T = ABC$ and its polar triangle $T_1 = A_1B_1C_1$. To find the side a_1 we extend (if necessary) the great-circle arc $\widehat{B_1C_1}$ so as to cut AC (or its extension) in D and to cut $\widehat{AB}$ (or its extension) in E. By the definition of polarity B_1 is the pole of $\widehat{AC}$ and C_1 the pole of $\widehat{AB}$. Therefore the arcs $\widehat{B_1D}$ and $\widehat{C_1E}$ are quadrants. These two arcs overlap on $\widehat{C_1B_1}$, which corresponds to the desired side a_1. But also, as we have shown, A is the pole of $\widehat{B_1C_1}$. Hence

$$\alpha = \sphericalangle \widehat{DE} = \sphericalangle \widehat{DB_1} + \sphericalangle \widehat{C_1E} - a_1 = \pi - a_1,$$

where $\sphericalangle \widehat{XY}$ is the angle at which the arc $\widehat{XY}$ appears at the center O of the sphere. Therefore $a_1 = \pi - \alpha$ and the same reasoning gives us the other two sides:

$$a_1 = \pi - \alpha, \qquad b_1 = \pi - \beta, \qquad c_1 = \pi - \gamma. \tag{19}$$

No further argument is needed to find the angles α_1, β_1, γ_1; since the relation of polarity is reciprocal, we get by applying (19) to the *polar* triangle T_1

$$\alpha_1 = \pi - a, \qquad \beta_1 = \pi - b, \qquad \gamma_1 = \pi - c. \tag{20}$$

In words, (19) and (20) state that the sides of T_1 are supplements of the corresponding angles of T, and its angles are the supplements of the sides of T.

The polar duality gives us a new theorem in spherical trigonometry for each theorem already proved. For instance, let us apply the spherical cosine law (13)–(15) to the *polar* triangle T_1. Using (19) and (20) we get

$$\cos \gamma = -\cos \alpha \cos \beta + \sin \alpha \sin \beta \cos c, \qquad (21)$$

$$\cos \beta = -\cos \alpha \cos \gamma + \sin \alpha \sin \gamma \cos b, \qquad (22)$$

$$\cos \alpha = -\cos \beta \cos \gamma + \sin \beta \sin \gamma \cos a, \qquad (23)$$

which is the dual law of spherical cosines. Similarly, applying the sine law (18) to the polar triangle, we have the identity

$$(1 - \cos^2 a - \cos^2 b - \cos^2 c + 2 \cos a \cos b \cos c)^{1/2}$$

$$(1 - \cos^2 \alpha - \cos^2 \beta - \cos^2 \gamma + 2 \cos \alpha \cos \beta \cos \gamma)^{1/2}$$

$$= \sin a \sin b \sin c \sin \alpha \sin \beta \sin \gamma.$$

We are now able to solve any spherical triangle T. First, let us enumerate the various cases that may arise, that is, the *different* ways in which three of the six elements of T may be given.

Three sides	Case 1	a, b, c
Two sides and an angle	Case 2	a, b, γ
	Case 3	a, b, α
Two angles and a side	Case 4	α, β, a
	Case 5	α, β, c
Three angles	Case 6	α, β, γ

However, only the cases 1, 2, and 3 need be considered; this is shown by exploiting the polar duality. By passing over to the spherical triangle polar to the one we wish to solve, it is found that case 6 is the exact polar dual to case 1, case 5 to case 2, and case 4 to case 3. Therefore all cases can be solved once we know how to handle 1, 2, and 3.

Solving spherical triangles is in general more complicated than solving the plane ones, at least in the sense that there are many special subcases because more things can "go wrong." We start by illustrating the types of such pathologies.

1. *Impossibility.* The three given elements do not determine any spherical triangle or, equivalently, the hypothetical spherical triangle with

the preassigned three elements does not "close up." A simple plane analog is the nonexistence of a plane triangle with sides a, b, c if $a > b + c$.

2.   Collapse.   The spherical triangle with the preassigned three elements collapses to an arc, i.e., an angle is 0 or π. The plane analog is when one side of a plane triangle is the sum of the other two.

3.   *Ambiguity.*   There are two distinct spherical triangles with the three given elements. This occurs when we are forced to use *sines*: For some unknown element x the value of $\sin x$ is unique but x itself has *two* possible values. These are supplementary: If x_1 is one value then $\pi - x_1$ is the other one. Note that such ambiguity does not arise when cosines alone are used: If $\cos x$ is known then x still has two possible values between 0 and 2π; these are x_1 and $2\pi - x_1$. Therefore only one value satisfies the standard convention that all elements are $< \pi$.

A simple plane analog is shown in Fig. 7.5*a*, where a plane triangle is constructed, given the sides a, b, and the angle α. If $b \sin \alpha < a < b$, then there are two such triangles. We check that the two angles β are supplementary: $\beta_2 = \pi - \beta_1$. This arises, analytically, from the use of

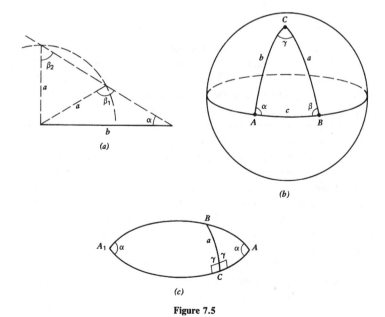

(a)

(b)

(c)

Figure 7.5

the plane sine law: β is given as $\sin \beta$ from

$$\frac{b}{\sin \beta} = \frac{a}{\sin \alpha}.$$

4. *Indeterminacy*. This pathology occurs when there are infinitely many triangles with the three given elements. Of the six elements of the spherical triangle several are right angles $\pi/2$. For that reason there is no plane analog. A simple example is shown in Fig. 7.5b. The desired spherical triangle ABC has two quadrants as its arcs: $a = b = \pi/2$. The opposite angles are also right: $\alpha = \beta = \pi/2$.

It is noted that if three elements out of the four a, b, α, β are given and are $\pi/2$, then the fourth one is also $\pi/2$, and so we have the preceding situation. The remaining two elements c and γ are arbitrary, and we have necessarily $c = \gamma$.

The following example considers a right-angled spherical triangle with $\gamma = \pi/2$. Since γ is known two more elements are needed, and we take these to be a and α; this constitutes case 4: One side is given and two angles, one adjacent and one opposite to the side. By the dual law of spherical cosines (23) we have

$$\cos \alpha = \sin \beta \cos a. \qquad (24)$$

Since $\sin \beta \leq 1$ and $\cos x$ is a decreasing function for $0 < x < \pi$, we must have $a \leq \alpha$. Therefore no spherical right-angled triangle exists with $\gamma = \pi/2$, a and α given, if $a > \alpha$. If $a < \alpha$, there are two such triangles; they are colunar, as is seen from the diagram of Fig. 7.5c. If the vertex C is free to move on the meridian AA_1, then it is clear that a is largest when C bisects the meridian AA_1 and then $a = \alpha$. In that case there is only one solution: triangles ABC and A_1BC are then congruent.

We consider now the basic problem of solving a spherical triangle. As was shown before, only cases 1, 2, and 3 need be considered. For case 1 the three sides a, b, c are given, where $0 < a$, b, $c < \pi$. Any three values are admissible provided that the triangle "closes up." The necessary and sufficient condition is the same as in the plane case, that is, of the three numbers a, b, c each is less than the sum of the other two. We solve the triangle by the spherical cosine law: Equations (13)–(15) enable us to find the angles by finding first $\cos \alpha$, $\cos \beta$, $\cos \gamma$. Since cosines alone are used there is no ambiguity and the solution is unique.

For case 2 any three values are admissible as a, b, γ, subject only to the general convention of being $<\pi$. Here the solution is also unique: First c is given by (13), then α and β are found from (14) and (15); again, the uniqueness of our solution follows, since only cosines are used.

It is case 3 that is the nasty one: to solve the spherical triangle when two sides a, b and the angle α opposite one of them are given. The *technique* alone of the solving process is simple enough. First, the angle β is found from the sine law (18):

$$\frac{\sin \alpha}{\sin a} = \frac{\sin \beta}{\sin b}$$

so that

$$\sin \beta = \sin \alpha \frac{\sin b}{\sin a} \tag{25}$$

The remaining pair of elements c and γ is still to be found. The cosine law (13) and the dual law (21) give us

$$\cos c - (\sin a \sin b) \cos \gamma = \cos a \cos b \tag{26}$$

$$-(\sin \alpha \sin \beta)\cos c + \cos \gamma = -\cos \alpha \cos \beta,$$

which is a system of two linear equations for the two unknowns $\cos c$ and $\cos \gamma$.

The difficulty comes when we observe that (25) or (26) may have sometimes no solution and sometimes more than one solution. Here it must be remembered that the sines and cosines are restricted to lie between -1 and 1. So, if the given elements a, b, α are such that

$$\sin \alpha \sin b > \sin a,$$

then there is no solution, for (25) shows that $\sin \beta > 1$, which is impossible. If $\sin \alpha \sin b = \sin a$, then by (25) $\sin \beta = 1$, and so there is unique value $\beta = \pi/2$. When

$$\sin \alpha \sin b < \sin a,$$

then (25) gives us a unique value for $\sin \beta$, which in turn gives us *two* admissible values for β. Therefore there are then two systems (26) for the unknowns $\cos c$ and $\cos \gamma$, corresponding to those two possible values of β. The following list gives a complete description of the different possibilities; there IMP stands for impossibility, IND for indeterminacy, COLL for collapse, UNIQ for uniqueness, and AMB for ambiguity.

1. $\alpha = \pi/2$ $b = \pi/2$ $a \neq \pi/2$ IMP

                            $a = \pi/2$ IND

            $b < \pi/2$ $a < b$ or $a > \pi - b$ IMP

                            $a = b$ or $a = \pi - b$ COLL

                            $b < a < \pi - b$ UNIQ

            $b > \pi/2$ $a < \pi - b$ or $a > b$ IMP

                            $a = b$ or $a = \pi - b$ COLL

                            $\pi - b < a < b$ UNIQ

2. $\alpha < \pi/2$ $b = \pi/2$ $a < \alpha$ or $a > \pi/2$ IMP

                            $a = \alpha$ UNIQ

                            $\alpha < a < \pi/2$ AMB

                            $a = \pi/2$ COLL

            $b < \pi/2$ $\sin a < \sin b \sin \alpha$ or $a > \pi - b$ IMP

                            $\sin a = \sin b \sin \alpha$ or $b \leq a < \pi - b$ UNIQ

                            $\sin a > \sin b \sin \alpha$ and $a < b$ AMB

                            $a = \pi - b$ COLL

            $b > \pi/2$ $\sin a < \sin b \sin \alpha$ or $a > b$ IMP

                            $\sin a = \sin b \sin \alpha$ or $\pi - b \leq a < b$ UNIQ

                            $\sin a > \sin b \sin \alpha$ and $a < \pi - b$ AMB

                            $a = b$ COLL

3. $\alpha > \pi/2$ The spherical triangle ABC with the elements a, b, α is colunar to AB_1C with the elements $\pi - a$, b, $\pi - \alpha$ so that $\pi - \alpha < \pi/2$. Now this case reduces to 2 above, with α replaced by $\pi - \alpha$ and a by $\pi - a$

To indicate some of the many applications of spherical trigonometry we end with two illustrative examples.

Example 1. Assuming spherical earth find the distance from city X of latitude θ_1 and longitude ϕ_1 to city Y of latitude θ_2 and longitude ϕ_2.

We start by drawing on the earth sphere the terrestrial triangle shown in Fig. 7.6a. This is the spherical triangle whose vertices are X, Y, and the north pole N. In this terrestrial triangle the two sides a and b and the included angle γ are known: $\gamma = |\phi_1 - \phi_2|$ if both longitudes are west of Greenwich or if both are east, and $\gamma = \phi_1 + \phi_2$ if one of them is west and the other east. The sides a and b are $90° -$ latitude, for north latitudes, and $90° +$ latitude for south latitudes. In our example X is in the Northern Hemisphere so that $a = 90° - \theta_1$ and Y is in the Southern

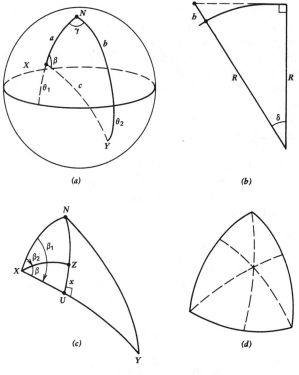

Figure 7.6

Hemisphere so that $b = 90° + \theta_2$. Now the side c is uniquely given by the cosine law (13). We find c, convert it to radians, and then the desired distance $\widehat{XY}$ is Rc, where R is the radius of earth. It may be also necessary to compute the bearing from X: this is the angle β giving the direction in which the motion from X to Y along $\widehat{XY}$ starts. We compute β from the cosine law (14).

Example 2. Let X and Y be as in the previous example and let city Z have latitude θ_3 and longitude ϕ_3. Find the distance of closest approach of Z to the arc $\widehat{XY}$.

This problem might arise as follows. An airplane flies from X to Y on the great-circle route $\widehat{XY}$ at constant altitude h; assuming spherical earth and perfect visibility, will Z be visible some time during the flight? From

Fig. 7.6b we find that the angle δ is given by sec $\delta = 1 + h/R$, where R is the radius of earth. Therefore visibility of Z occurs if and only if the minimum angular distance of Z to $\widehat{XY}$ is $<\delta$. A grimmer possibility is when an enemy missile is shot from X at Y along the great-circle route $\widehat{XY}$, the defense countermissile is stationed at Z and has a given range, and we want to know whether the countermissile can reach, and destroy, the missile.

Suppose that Z is in the position shown in Fig. 7.6c so that the point U of closest approach of Z to $\widehat{XY}$ lies as shown on $\widehat{XY}$. By using Example 1 with the terrestrial triangle XZN we compute the side c and the angle β_2. Then from the terrestrial triangle XYN we compute the angle β_1. Now we have the right-angled triangle XZU in which the hypotenuse c and the angle $\beta = \beta_1 - \beta_2$ are known. Therefore the side x corresponding to $\widehat{ZU}$ can be found by the sine law:

$$\frac{\sin \beta}{\sin x} = \frac{\sin 90°}{\sin c} = \frac{1}{\sin c},$$

and so sin $x = \sin \beta \sin c$. Finally, the distance $\widehat{ZU}$ is xR. Attention must be paid to the possible ambiguity of x once sin x is known.

EXERCISES

1. On the axial-projection diagram of Fig. 7.1a let K be a great circle of S whose plane is neither vertical nor horizontal. Describe the axial projection of K onto the cylinder C.

2. The province of Personitoba extends exactly from latitude 48° north to latitude 55° north and from longitude 76° west to longitude 83° west. Assume spherical earth and find the area of Personitoba.

3. A spherical polygon is bounded by n arcs of great circles on a sphere of radius r. If its angles are $\alpha_1, \alpha_2, \ldots, \alpha_n$, find its area.

4. What is the steradian content of a cone whose semivertical angle is α?

5. Prove the spherical cosine law (13) when one or both sides a, b are $\pi/2$.

6. Deduce the sine law $a/\sin \alpha = b/\sin \beta = c/\sin \gamma = 2r$ for a plane triangle with sides a, b, c, angles α, β, γ, and circumradius r, from

the spherical sine law (18). (*Hint*: Write a/R, b/R, c/R for a, b, c in (18) and use the expansions $\sin x = x - \ldots$, $\cos x = 1 - x^2/2 + x^4/24 - \ldots$, then let $R \to \infty$.)

7. Let T be a right-angled spherical triangle with $\gamma = \pi/2$. Show that $\tan a = \tan \alpha \sin b$.

CHAPTER EIGHT
INTRODUCTION
TO GRAPHS

In this section we introduce new kinds of geometrical objects called *graphs* and we show by means of examples several different ways in which such graphs may turn up in mathematical practice. In simple terms, a graph consists of a finite number of points v_1, v_2, . . . , v_n called its *vertices* and a finite number of arcs e_1, e_2, . . . , e_p called its *edges*. Each edge joins a vertex to another vertex or, possibly, to itself. Infinite graphs, where the numbers of vertices or edges are infinite, could also be considered but we shall not do so: All our graphs will be *finite*.

To give at this early stage some idea of the possibilities of graphs and to motivate some later terminology, we consider the following "problem of six people": Show that among any six people there must be either some three of whom every two are acquainted, or else, some three of whom every two are strangers.

The first thing to do is to ask whether five people would also do instead of six. The answer is no: Six is the best, that is, the smallest possible number. This is proved by giving a suitable counterexample of five people. Let us consider the five vertices v_1, v_2, v_3, v_4, v_5 of Fig. 8.1a, which stand for the five people, and the "acquaintanceship graph" of five solid edges, in which two vertices are joined by an edge if and only if they are acquainted. The dotted lines are the edges of the "strangeness graph" in which two vertices are joined by an edge if and only if they are strangers. For convenience of expression let us pretend that the solid edges of the acquaintanceship graph are black whereas the dotted ones of the strangeness graph are red. As Fig. 8.1a shows, there is no monochromatic triangle, that is, no triangle with all three sides of the same color. This shows that the number 6 in our problem cannot be reduced to 5. The graphs of Fig. 8.1a express the relationship of five people sitting around a small circular table, if every two neighbors know each other but no others do.

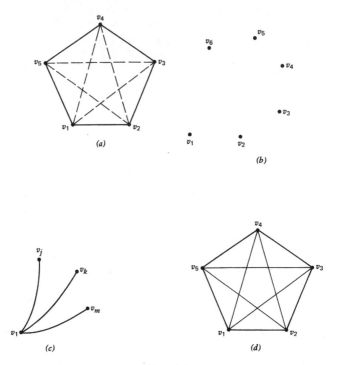

Figure 8.1

Suppose next that there are six vertices, for the six people as in Fig. 8.1b, and consider the vertex v_1. To express its relationship to the other five, there are the five arcs v_1v_i, $i = 2, 3, 4, 5, 6$, each of which must be either black for acquaintance or red for strangeness. Since $2 + 2 < 5$, it follows that of the five edges v_1v_i at least three must be of one color, as is shown in Fig. 8.1c. Let these three edges be v_1v_j, v_1v_k, v_1v_m; it is irrelevant whether they are black or red, for by symmetry the argument works in either case. Consider finally the vertices v_j, v_k, v_m of Fig. 8.1c and the triangle of the three edges joining those three vertices; this triangle is not shown in Fig. 8.1c. If the triangle is monochromatic, we have proved what we want. If it is not, then it has edges of both colors, and so one edge must be of the same color as the three edges from v_1. Therefore it completes certain two of them to a monochromatic triangle, and so again we are finished.

We see that is quite irrelevant whether the edges are represented as straight or not: Only the *connectivity* properties interest us here, and not the metric properties, such as the edge lengths. Two edges may cross in the graph, as the dotted lines do in Fig. 8.1*a* or the edges in Fig. 8.1*d*, and their intersection point is *not* a vertex of the graph. Observe that any acquaintanceship or strangeness graph must by its very nature have the following two properties:

1. There are no *loops*, that is, no edges joining a vertex to itself.
2. There are no *multiple* edges, that is, no two edges joining the same two vertices.

Such graphs occur very often and there is a name for them: a graph satisfying properties 1 and 2 is called *special*. If either loops or multiple edges are allowed, the graph is called *general* or, occasionally, a *multigraph*. Often the qualification *special* or *general* is left out when it is clear in the specific situation which type of graph is meant.

If in Fig. 8.1*a* the color of edges is disregarded, we obtain what is known as the *complete* graph with, or on, five vertices. This is shown in Fig. 8.1*d*: Every two of its five vertices are connected by an edge. In analogy to this complete 5-graph we can define a complete *n*-graph for every $n > 1$. For instance, a triangle is a complete 3-graph and a square together with its two diagonals is a complete 4-graph. Note that the four vertices and the six edges of a tetrahedron also form a complete 4-graph.

It is clear that any acquaintanceship graph and its matching strangeness graph are *complementary*: (1) they have the same *n* vertices, (2) they have no edges in common, (3) together they make up the complete *n*-graph. Thus, to any graph G on *n* vertices there corresponds a unique *complementary graph* $\bar{G}$: It has the same vertices as G but only those edges of the complete *n*-graph that are missing from G.

With this terminology the problem of six people can be restated technically: If G is any graph on six vertices, then either G or $\bar{G}$ contains a complete 3-graph. This restatement leads at once to the following generalization, known as Ramsey's theorem:

For every $n > 1$ there is a finite number $f(n)$ with the following property: if G is any graph on $f(n)$, or more, vertices, then either G or $\bar{G}$ contains a complete *n*-graph.

We could prove it here but we shall not do so since the subject belongs to combinatorics rather than to geometry. It is of interest to observe that

$f(3) = 6$ as the problem of six people says, further $f(4) = 18$, but the (smallest) value of $f(n)$ is not known for any $n > 4$. However, it has been proved that $f(n) \leq \binom{2n-2}{n-1}$. F. P. Ramsey (1903–1930), after whom the theorem is named, was an English philosopher and mathematician who died at the age of 26 of smallpox; his brother Michael was for many years the archbishop of Canterbury.

If G is any acquaintanceship graph and v is a vertex of G, we may be interested in such things as all people who know v, then all people who know people who know v, etc. This suggests the following definitions. In a graph G all vertices that send edges to v are *adjacent* to it, and their number is the *degree* (or *weight*, or *valency*) of v. So, in an acquaintanceship graph, all vertices adjacent to v represent the people whom v knows, and degree of v tells us how many people v knows. A *link* or a *path* from a vertex v_1 to a vertex v_2 is a chain of successive edges $v_1 v_a$, $v_a v_b, \ldots, v_z v_2$ that starts at v_1 and terminates at v_2. For a graph G such a link may arise from the study of *spreading*, for instance, of rumors or of infection, along a chain of successively adjacent vertices. How far can such things spread, starting from a vertex v? A graph G is *connected* if and only if every two of its vertices can be joined by a link. If G is not connected, then it decomposes into two or more *components*; here a *component* is a maximal connected subgraph. That is, a component G_1, of a graph G is a subgraph in which every two vertices can be linked and which cannot be extended to a larger subgraph in which such linking occurs. Now our question is simply answered: In a graph G a rumor or an infection starting at a vertex v can spread at most to that component of G which contains v. If G is connected, then the sole component is G itself. A closed link, which starts and ends up at the same vertex, is called a *cycle*; it is a generalization of a loop.

Our next problem is the historically famous and important one of the "seven bridges of Königsberg," solved in 1736 by the great Swiss mathematician Leonhard Euler (1707–1783). This concerns the bridges on the river Pregel at Königsberg, in East Prussia. This city was once Polish, under the name Królewiec, then it became a property of Germany, as Königsberg; currently it is Russian, rebaptized as Kaliningrad. The two banks of the river and its two islands are connected by seven bridges shown in Fig. 8.2a. The citizens of Königsberg wanted to know whether it was possible to make a walking tour in which every one of the seven bridges was crossed exactly once. Stripped of its irrelevancies, the problem is illustrated in Fig. 8.2b by the graph showing the connectivity of

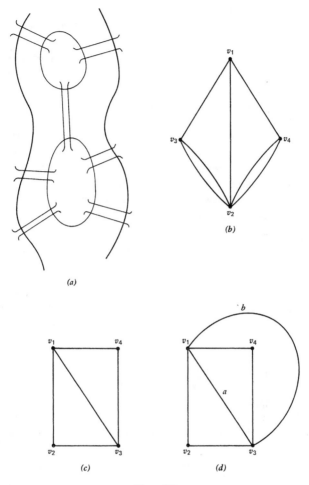

Figure 8.2

the four land masses. There are four vertices: v_3 and v_4 represent the left and the right bank, v_1 and v_2 represent the two islands; the edges are the connecting bridges. There are no loops but multiple edges do exist, thus we deal here with a general graph; also, this graph is connected. It is noted that v_1, v_3, and v_4 have degree 3, whereas v_2 has degree 5. Therefore the sum of all degrees is 14, an even number.

This is no accident. The sum S of all degrees is even in any graph G. For, what does S count? The answer is that it counts the total number of edges, but every edge is counted *twice*. Each edge is counted once from each end, and if this edge happens to be a loop, it is still counted twice. So, since S is twice the number of edges, it must be even. For instance, the sum 14 of degrees in Fig. 8.2b is twice the number of edges, which is 7—the seven bridges.

A simple parity argument shows further that the number of odd-degree vertices must always be even. This is seen as follows. The contribution to S from even-degree vertices is itself even; therefore the contribution from odd-degree vertices is also even since the sum of the two contributions is even. Hence the number of odd-degree vertices could not possibly be odd. It follows that the number of odd-degree vertices is 0, 2, 4, ... For instance, the number of such odd-degree vertices in Fig. 8.2b is 4, in Fig. 8.2c it is 2, and in Fig. 8.2d it is 0.

We define now an Euler path and an Euler circuit for a graph G; these concepts arose precisely out of Euler's solution of the traversing problem for the seven bridges of Königsberg. An *Euler path* of G is a link that crosses every edge of G exactly once. For instance, the graph of Fig. 8.2c has the Euler path $v_1v_2 - v_2v_3 - v_3v_1 - v_1v_4 - v_4v_3$. As we see, every edge is visited once only, though a vertex, such as v_1, may be visited several times. An *Euler circuit* is a particular Euler path that is a cycle, starting and ending at the same vertex. For instance, the graph of Fig. 8.2d admits the Euler circuit $v_1v_2 - v_2v_3 - v_3(a)v_1 - v_1v_4 - v_4v_3 - v_3(b)v_1$. We have now everything needed to prove the principal result, which is the following theorem of Euler:

(E)   A graph G has an Euler path if and only if it is connected and has either zero or two odd-degree vertices; G has an Euler circuit if and only if it is connected and there are no odd-degree vertices.

This disposes effectively of the problem of the seven bridges of Königsberg: The graph of Fig. 8.2b has four vertices of odd degree, and so there is no Euler path. Hence a walking tour crossing each bridge once is impossible.

In proving (E) no distinction need be made between showing the existence of an Euler path and showing the existence of an Euler circuit; each implies the other. The case of the Euler circuit is reduced to the Euler path by removing one edge from G. The removed edge joins two vertices, say a and b, which now acquire odd degrees. The Euler path from a to b exists, by hypothesis, and it is now completed to an Euler circuit by returning the removed edge. Conversely, the Euler path is changed to an Euler circuit by adding an edge; it is the edge joining the two odd-degree vertices, and so now all vertices have even degrees. By hypothesis, an Euler circuit exists; we now remove the added edge, and so on.

Suppose now that an Euler path, or circuit, E exists for G. It is then obvious that G is connected. Every time E crosses a vertex v of G it passes over exactly two edges from v, excepting only the first and the last vertex for the Euler path, and with no exceptions for the Euler circuit. Thus the degree of every vertex is even, except that for the Euler path the first and the last vertices have odd degrees. This proves the necessity in (E).

Sufficiency is proved by induction on the number of edges of G. There is no difficulty to start the induction by proving that (E) holds for all graphs with two, or three, or four edges. Suppose now that G has two vertices a, b of odd degree and is connected, and assume that the existence of an Euler path from a to b has been proved whenever G has $\leq n$ edges. Suppose next that G has $n + 1$ edges. We produce first the longest path E_1 from a to b in G. This is the path using the greatest possible number of edges, and each of them once only. If E_1 uses all the edges, it is an Euler path and we are finished. If not, remove from G all the edges of E_1, and let G_1 be the remaining graph. The reasoning used before with the crossing of vertices shows that every vertex of G_1 has an even degree. Since G is connected, G_1 must have a vertex v on E_1; let G_{11} be the component at G_1 containing v. Then, *by the induction hypothesis*, G_{11} has an Euler circuit. But now this Euler circuit can be spliced together with E_1 by a juncture at v. However, this contradicts the definition of E_1 as the *longest* path. Hence the induction is complete and (E) is proved.

Just as the Eulerian paths and circuits visit every edge of a graph G exactly once, so the *Hamiltonian* paths and circuits visit every vertex of G exactly once. It is easily proved that a Hamiltonian path crosses every edge of G at most once, though it need not cross any particular edge. The subject of Hamiltonian paths is much harder than that of Eulerian paths: No condition is known to be necessary and sufficient for a graph to have a Hamiltonian path. However, some partial results in this direction exist. The Hamiltonian paths and circuits were introduced by Sir W. R. Hamilton (1805–1865), an Irish mathematician, in the form of a game related to the regular dodecahedron D. In this game the 20 vertices of D are considered to be cities; it is asked whether a round trip on the 30 edges of D exists that visits every city exactly once. As is shown in Fig. 8.3, the answer is yes.

The next problem considered goes back to the ancient Greek legend of Theseus and Ariadne. Theseus was an Athenian who traveled to Crete and entered there a famous maze-building called the labyrinth. His purpose was to find the monster Minotaur who resided somewhere in the labyrinth, kill him, and then find his way out of the maze. Theseus was aided in his task by the Cretan princess Ariadne and her gift of a ball of thread. This was tied up at the entrance, unrolled as he went in, and then rolled up when he returned. The legend goes on to say that Theseus, on his way back to Athens, left Adiadne stranded on the isle of Naxos (where she died in childbirth).

Even the preceding schematic description shows that Theseus' problem is (in part) another question of traversing a graph. The maze of the labyrinth is represented by a connected graph; an example of a small labyrinth is shown in Fig. 8.4. The edges are marked by small letters and they represent the corridors, or passages, of the labyrinth. The vertices are marked by capital letters, and they represent the junctions where the corridors or passages meet. There may be blind coridors leading nowhere, like s; the dead-end junction G is then a vertex of degree 1. Undercrossings and overcrossings may occur, as with edges r and q. There are loops like

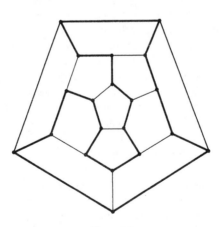

Figure 8.3

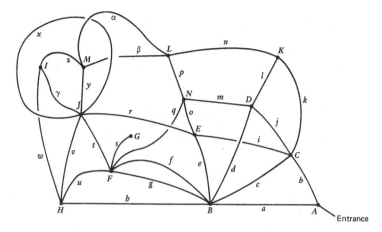

Figure 8.4

x, and multiple edges, like f and g. Thus the graph of the labyrinth is a general graph.

One vertex, A in Fig. 8.4, is marked as the entrance, and also exit; another vertex, M in Fig. 8.4, is the fixed residence of the Minotaur. The conditions of search suppose complete ignorance of the plan of the labyrinth and total inability to recognize an edge or a vertex which has been visited. We might imagine that Theseus moves either blindfolded or in total darkness. It is supposed that, feeling his way along the walls or otherwise, Theseus can recognize a junction when he comes to one and can count all the corridors meeting there. It will be shown that with these three abilities—to recognize junctions, to count the edges, and to remember—Theseus does not need Ariadne. That is, a method exists that enables him to enter any finite labyrinth, find any fixed object (i.e., vertex in it), and then return to the entrance. It is emphasized that the method must work for any finite labyrinth, so long as it is connected.

We start Theseus in the entranceway facing A. Starting from the right-hand corridor b, Theseus enumerates all the edges meeting at A. There are two of them and so b is 1 and a is 2. Suppose that edge 2 is picked, that is, the edge marked a. Theseus moves on along it to the next junction, which is B. Standing in a facing B, he repeats the enumeration process. Vertex B has degree 7; thus c is 1, d is 2, e is 3, f is 4, g is 5, h is 6, and a itself is 7. Let Theseus pick again a corridor, for instance, f, which has number 4. He moves along to the junction F, where the procedure is repeated. Suppose that from F he chose corridor 2, that is, s. He gets then to the dead-end junction G; then he chooses the only way out, i.e., returns to F. Next, he picks third corridor, that is, g, which takes him to B, and then the corridor 5 (from the right, as always) which is e; this lands him at the vertex E. His travels are shown in the following table:

At junction	Took corridor no. (from right)	Which is
A	2	a
B	4	f
F	2	s
G	1	s
F	3	g
B	5	e
E		

Theseus memorizes the number sequence 2, 4, 2, 1, 3, 5 that codes this particular trip. Now suppose that at this particular time, being at E, he decides to return to the entrance A.

Standing in the corridor e and facing E, he turns round and returns to the vertex B. Here he takes corridor 5 but *counting from the left*. This is g, which takes him back to F. Here he takes corridor 3, again from the left, and this is s, which takes him to the dead-end G. He continues in the same way; by using the memorized numbers in reverse, as 5, 3, 1, 2, 4, 2 and always counting from the left, he arrives at his starting point which is the entrance A. In this way it is possible to return to A after any trip coded by the sequence $n_1, n_2, \ldots, n_a$, by reversing the sequence, and counting from the left.

Now, starting at A, Theseus executes all possible trips of length 1 returning to A after each, then all possible trips of length 2, of length 3, and so on. Since Theseus can count the degree of each vertex, i.e., the number of corridors meeting there, there is no difficulty in checking off all possible trips of any given length. Eventually Theseus arrives at the vertex M, settles his business with Minotaur, and then returns to A.

The strategy of his search becomes clearer if we produce for him the following diagram of Fig. 8.5, which forms the pattern of search. This starts from the initial vetex A, the entrance, from which emerge two edges marked 1 and 2, and leading down to vertices C and B. From each of these there starts the correct number of further downward edges leading

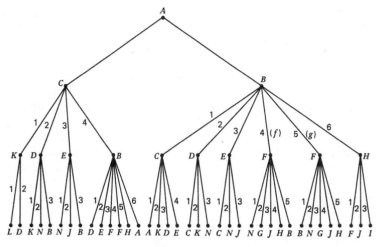

Figure 8.5

to new vertices, and so on. We note that this search tree, as it is called, is itself a special kind of a graph, together with a preferential ordering. This preference proceeds from the enumeration of corridors starting from the first one to the right. Theseus' trips of length 1 search the tree down to level 1, that is, to the level of C and B; trips of length 2 take him down to level 2 with the vertices K, D, E, B, C, D, E, F, H, and so on. Figure 8.5 shows all trips of length 3, that is, shows the complete pattern of Theseus' search down to depth 3. If there were room to show one more level, to depth 4, then the vertex M would already appear (several times, too). So Theseus need not extend his labyrinth search below depth 4. In particular, the sample trip we have considered is coded as 242135 and takes Theseus down the search tree to depth 6. Of course, this trip would not be executed for our particular labyrinth because the Minotaur M is found at depth 4.

It is clear that the method will always work. But the size of the labyrinth is not known in advance and so Theseus may need very good memory. That is to say, he may be forced to memorize arbitrarily long strings of numbers. For this reason the labyrinth search problem may be called unbounded, i.e., calling for unbounded memory. Is it possible to change the conditions so as to execute the search without having to memorize arbitrarily much information?

Suppose that Theseus goes forth "by day" and can see the edges through which he passes. Let him be given Ariadne's ball of thread which he can unroll as he goes on, or rewind if he decides to backtrack. Finally, let us suppose that Theseus has some means of marking a corridor that he crosses, if he so wishes. Under these conditions it may be shown that Theseus can find M, and then return to A, by means of following four simple instructions, and without any other demands on his memory.

Theseus begins by tying his thread at the entrance vertex A and he moves along corridors, or edges, from vertex to vertex. A corridor is called *new* if it neither is marked nor has thread running through it. The *backtracking* operation is as follows: At some vertex v Theseus returns along the edge by which he came to v, rewinding the thread as he returns, and then he *marks* the edge he has just crossed. The four instructions are as follows:

If the new vertex v:	Then Theseus does this:
Is M	Does his job and returns along the path of the thread
Has the thread through it	Backtracks
Has no thread through it but also no new corridor from it	Backtracks
Has no thread through it and has a new corridor from it	Unrolls the thread along any new corridor from v

The initial vertex A always counts as having the thread through it. It is seen that Theseus never crosses any corridor more than twice. Hence he cannot get lost or "start moving in circles." Since the whole labyrinth is finite and connected, there is a simple path from A to M and, sooner or later, Theseus finds it.

The next problem considered is completely different: how to cut up a rectangle R into a finite number of smaller rectangles $R_1, R_2, \ldots, R_n$. With every such decomposition of R there goes a certain graph, that shows which rectangles R_i lie immediately above which other ones. It will turn out that this graph, considered as an electrical network, has certain interesting special properties.

The basic rectangle R is taken in the standard orientation as in Fig. 8.6a. The smaller rectangles into which R gets decomposed do not overlap except possibly for sharing parts of their boundaries; since their number is finite, it follows that they are in the same orientation as R and cannot

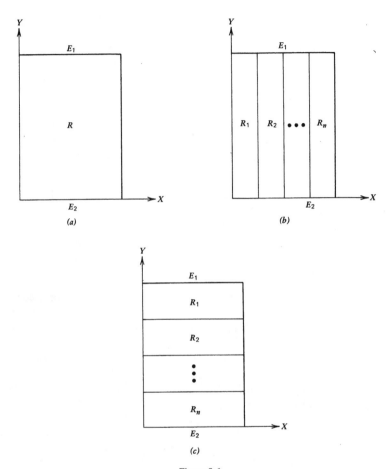

Figure 8.6

lie obliquely with respect to R. Let us associate the vertical dimension of rectangles with voltage, and their horizontal dimension with current. In this way a rectangle R_i of height a and base b gets identified with the electrical resistance $R_i = a/b$.

Consider first the two simple decompositions of R shown in Fig. 8.6b and c. Let R have the height a and base b, and in Fig. 8.6b let R_i have the base b_i so that $b_1 + b_2 + \cdots + b_n = b$. Each R_i has the same height a as R. Let us now produce the vertical adjacency graph for Fig. 8.6b.

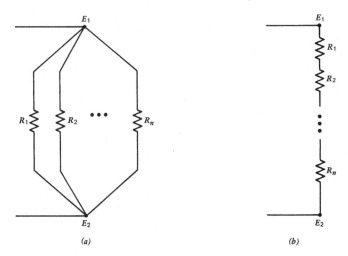

Figure 8.7

We have the upper vertex E_1 in Fig. 8.7a corresponding to the upper edge of R in Fig. 8.6a, and similarly the lower vertex E_2. Since no rectangle lies above or below any other, the vertical adjacency graph is as shown in Fig. 8.7a: The n resistances $R_1, R_2, \ldots, R_n$ are in parallel connection. In Fig. 8.6c the n rectangles are completely stacked up one over another, and so the vertical adjacency graph is as in Fig. 8.7b: The n resistances $R_1, R_2, \ldots, R_n$ are in series connection. We know that the area of R is ab; therefore for Figs. 8.6b and 8.7a

$$ab = a(b_1 + b_2 + \cdots + b_n). \qquad (1)$$

The assignment of electrical resistances in Fig. 8.7a is

$$R = \frac{a}{b}, \qquad R_1 = \frac{a}{b_1}, \qquad R_2 = \frac{a}{b_2}, \qquad \ldots, \qquad R_n = \frac{a}{b_n};$$

dividing (1) by a^2 we find that

$$\frac{1}{R} = \frac{1}{R_1} + \frac{1}{R_2} + \cdots + \frac{1}{R_n}$$

which is the correct expression for the net resistance R of n resistances $R_1, R_2, \ldots, R_n$ in parallel. Similarly, for Figs. 8.6c and 8.7b we have

$$ab = (a_1 + a_2 + \cdots + a_n)b, \qquad (2)$$

where $a_1, a_2, \ldots, a_n$ are the heights of horizontal rectangles $R_1, R_2,$ $\ldots, R_n$ and b is their common width. The assignment of electrical resistances in Fig. 8.7b is

$$R = \frac{a}{b}, \qquad R_1 = \frac{a_1}{b}, \qquad R_2 = \frac{a_2}{b}, \qquad \ldots, \qquad R_n = \frac{a_n}{b},$$

and so, dividing (2) by b^2, we find that

$$R = R_1 + R_2 + \cdots + R_n,$$

which is the correct expression for combining resistances in series.

We take up next a different decomposition of the basic rectangle R, as shown in Fig. 8.8a, into five rectangles: R_1, R_2, R_3, R_4, R_5. In addition to the upper and lower edges E_1 and E_2 we label also the two intermediate

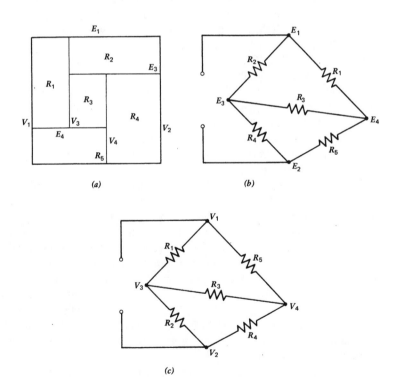

(a)   (b)

(c)

Figure 8.8

horizontal edges E_3 and E_4. How do we produce the vertical adjacency graph for this case?

We start with the upper terminal E_1 of Fig. 8.8b corresponding to the upper edge of Fig. 8.8a. Two rectangles, R_1 and R_2, lie below E_1 in Fig. 8.8a. Therefore two resistances R_1 and R_2 start down from the terminal E_1 of Fig. 8.8b: R_1 goes to E_4 and R_2 goes to E_3. Similarly, the rectangle R_3 lies below E_3 and above E_4 in Fig. 8.8a, and so the resistance R_3 goes down from E_3 to E_4 in Fig. 8.8b. Finally, R_4 goes from E_3 down to E_2, and R_5 from E_4 down to E_2. We have therefore, and quite unexpectedly, obtained the bridge network of Fig. 8.8b.

The rectangular decomposition of Fig. 8.8a also gives rise to another electrical network by means of horizontal rather than vertical adjacency graph. This tells us which rectangle is next to which, rather than above which one. The vertical edges V_1 and V_2 are labeled as in Fig. 8.8a, and also the two intermediate vertical edges V_3 and V_4. Proceeding just as before, we start from the terminal V_1 of Fig. 8.8c and we get then another electrical network, which is shown in Fig. 8.8c. It is dual in a certain sense to that of Fig. 8.8b. Both happen to be bridge networks, though with resistances placed differently.

We should now verify that the total resistances of the two networks of Fig. 8.8b and c are correctly given in terms of the height and base of R. Instead of this we shall produce a general verification: Any rectangular decomposition of the rectangle R leads by means of the adjacency graph to a two-terminal network of resistances, whose total resistance is again the ratio: height of R to base of R.

Take the basic rectangle R with its upper and lower edges E_1 and E_2. Let it be decomposed into a finite number of rectangles $R_1, R_2, \ldots, R_n$. Suppose that a potential is applied to the upper edge E_1 while E_2 is grounded. We suppose that R is electrically uniform, with some specific resistance. On account of the homogeneous uniformity there are no lateral currents, the whole flow of current being in the vertical direction down. That is, the points of any horizontal segment in R are electrically at the same potential. It follows that we can make in R all the *vertical* cuts of the subdivision into $R_1, R_2, \ldots, R_n$ without changing anything electrically. Finally, all the points on any horizontal single cut, or edge, can be lumped together into a terminal of our network, because they are at the same electrical potential.

It is now clear that with proper units the vertical dimension, or height, of any rectangle R_i represents the potential difference across the coresponding resistance R_i. At the same time, the horizontal dimension, or base, of the rectangle R_i represents the current through the resistance R_i. But the same is true for the whole of R as well. Hence the electrical resistance of the network consisting of resistances $R_1, R_2, \ldots, R_n$ connected as in the adjacency graph is correctly given.

From now on we shall limit ourselves to the case when R is a square and the constituent rectangles $R_1, \ldots, R_n$ are also squares. An example of such a "squaring of the square" is given in Fig. 8.9a, where the square R of side 23 is decomposed into 13 squares $R_1, \ldots, R_{13}$ of sides shown for each square. By means of the vertical adjacency we obtain the rather complicated network of Fig. 8.9b. This consists of 13 one-unit resistances; the total resistance of the network is also one unit, since for a square the height, or voltage, is numerically equal to the base, or current. Yet we note that our network is far from being of the simple series–parallel type, and its total resistance would not be easily found by any other method.

To avoid certain trivial square decompositions of a square we could place two further conditions on the constituent squares $R_1, \ldots, R_n$: (1) no subcollection of them makes up a rectangle, (2) no two squares are of the same size. If condition 1 is satisfied, the decomposition is called *simple*; if condition 2 is satisfied, it is called *perfect*. Thus our decompsotion of Fig. 8.9a is simple but not perfect, since our example has several pairs of equal-sized squares.

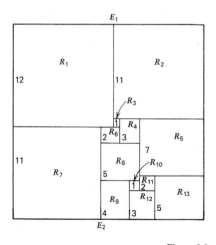 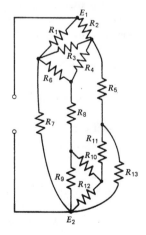

Figure 8.9

The problems now arise of the smallest number n of squares in a decomposition that is (1) simple, (2) perfect, (3) simple and perfect. As a matter of fact, it was believed that no decomposition both simple and perfect existed. This was contradicted by an example of a simple and perfect decomposition of a square into 38 squares. This value was later reduced and the best current figure appears to be 24 squares. We finish this topic by observing that it has a curious and very practical application. In building electrical, and particularly electronic, apparatus the need arises for many different resistances. But a manufacturer of electrical resistances will ordinarily supply only certain *standardized* values, say 1 ohm, 5 ohm, 10 ohm, and so on. Therefore there arises the problem of finding simple combinations of the standard resistances that use possibly few of them and provide the desired total values. Here our topic is of obvious use.

We take up now our last graph-theoretic topic; this is the Steiner problem mentioned earlier in the section on geometrical extrema. It is as follows: Given n points $A_1, A_2, \ldots, A_n$ in the plane, find the shortest network connecting them. There are obvious practical applications to various communication nets of roads, cables, canals, and so on. In particular, the problem arose in connection with the so-called videophone, a combination of television and telephone that allows one not only to speak but also to see at a distance. As can be imagined, the cable that carries both the voice and the view is much more complex and therefore also more expensive than ordinary telephone cable. This places a special premium on minimizing the total length of the net. We note that our minimal length network contains the given n points $A_1, A_2, \ldots, A_n$ as vertices but it may also contain any number of additional ones. It is this freedom of adding extra vertices that makes our problem difficult (but also interesting). It will always be assumed that all vertices, edges, graphs, and so on, lie in the plane.

First, we state our problem precisely, using the graph terminology developed earlier. To begin with, there are only the n given points $A_1, A_2, \ldots, A_n$; it is required to find

1. The integer k.
2. k additional points $S_1, S_2, \ldots, S_k$.
3. A connected graph G with the $n + k$ vertices $A_1, A_2, \ldots, A_n$, $S_1, S_2, \ldots, S_k$ whose total length $L(G)$ is minimum.

There is no difficulty about defining $L(G)$: it is the sum of all the edge lengths in G. Since each edge in G connects two points it must be a straight segment. Further, it is obvious that the minimal graph G has no loops since they serve here no useful purpose whatever. Their absence does; it suggests that our minimal graph G might perhaps have no cycles as well as no loops. This motivates the following definition:

(D_1) A *tree* is a connected graph without cycles.

There are two other possible definitions:

(D_2) A *tree* is a connected graph that becomes disconnected by removing any single edge (or briefly, a *tree* is a minimally connected graph).

(D_3) A *tree* is a graph in which every two vertices can be joined by one link only.

Which definition would we adopt? It does not matter: All three are equivalent. This is shown by proving the implications

$$D_1 \to D_3 \to D_2 \to D_1.$$

First, to show that $D_1 \to D_3$ is the same as to show that not-$D_3 \to$ not-D_1. But this is simple: If two different links join a pair of vertices and we erase their common edges (if any), then a cycle is produced. It is worthwhile to point out that the same reasoning is used in vector calculus when we prove that a line integral of a function is path independent if and only if any loop integral vanishes. The next implication $D_3 \to D_2$ is a simple consequence of the definitions D_3 and D_2 themselves: Since any two vertices can be joined by a link, the graph is connected; since the link is unique, the graph is minimally connected. Finally $D_2 \to D_1$ because not-$D_1 \to$ not-D_2: If there is a cycle in a connected graph and an edge of that cycle is removed, then the graph remains connected; this contradicts the minimal connectedness.

We return now to our minimal graph G; it obviously satisfies the definition D_2 because if it is possible to remove an edge, then it is possible to save on total length $L(G)$. Therefore the minimal graph G is a tree. This allows us to rephrase our whole problem very simply:

to find the shortest tree whose vertices include n given points A_1, A_2, $\ldots, A_n$.

Far more important, knowing that G is a tree will enable us to get a crucial upper bound on the number k of extra vertices that may be present. First, we prove a general proposition valid for any tree T: the number of its vertices exceeds the number of its edges by 1:

$$\text{number of vertices } - \text{ number of edges } = 1. \tag{3}$$

Start from any vertex v_1 of T and move along an edge to a neighboring vertex v_2, then along an edge from v_2 to a vertex v_3, other than v_1, and so on. Since T is finite, we either must return to a vertex already visited or else we run into a dead-end vertex: one of degree 1, that is, with only one edge. But a return to an already visited vertex is impossible since there are no cycles T; therefore there must exist a dead-end vertex v. Remove from T both the vertex v and its single edge; the result is a smaller tree on which the same operation is repeated. Eventually, from this process, a tree is produced that consists of just two vertices and the one edge joining them; here (3) obviously holds. Therefore it also holds for T, since the removal operations clearly do not change the L.H.S. of (3).

For the next result we recall Problem 6 of Chapter 6 and we make the following definition. Let v be a vertex of our minimal tree G such that $d(v) = 3$: We call v a Steiner vertex if and only if the three edges of v make angles 120° in pairs. Let u be any vertex of G, which may be either one of the original n points A_i or an additional vertex S_j; then

$$d(u) \le 3 \text{ and if } d(u) = 3, \text{ then } u \text{ is a Steiner vertex.} \tag{4}$$

To prove that $d(u) \le 3$ we suppose the contrary: Let u be a vertex with four edges from it (or more), as shown in Fig. 8.10a. Of course, the dotted lines in this figure are *not* edges of G. Now in each of the four triangles of the figure the dotted line must be at least as long as the two solid lines. For, otherwise, remove the longer (solid) side from G and replace it by the shorter (dotted) side. The result is another tree of shorter total length; this contradicts the minimality of G, and so is impossible.

Next we observe that in any triangle whatever the longer of any two sides lies opposite the larger of the two angles. It follows that for each of the four triangles of Fig. 8.10a the vertex angle at u is the largest angle of that triangle (i.e., it is larger than or equal to any other angle). Since the four vertex angles at u add up to 360°, the smallest of them is $\le 90°$. This means that at least one of the four triangles is nonobtuse (right or acute). Let it be the one shown in Fig. 8.10b. We now use Problem 6 of Chapter 6: Inside that triangle there is its Steiner point S, which minimizes

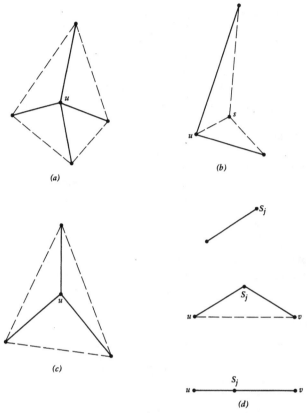

(a)

(b)

(c)

(d)

Figure 8.10

the sum of the three vertex distances. Therefore, in the triangle of Fig. 8.10*b* we erase the two solid sides and we add *S* to *G* as a new additional vertex, together with its three dotted sides as edges. The result is a graph shorter than *G*, which is a contradiction. Hence indeed $d(u) \leq 3$.

Suppose next that $d(u) = 3$ as in Fig. 8.10*c*. If all three angles at *u* are 120°, *u* is a Steiner vertex and we are finished. If not, then one of the three angles at *u* is <120°. Then for the triangle corresponding to that angle we use the same reasoning as in Fig. 8.10*a* and *b*: By erasing the two solid-line sides and introducing instead of them a new additional vertex *S* together with its three edges, we shall actually save on total length, which is a contradiction. This completes the proof of (4).

It is easily shown that each additional vertex S_j of the minimal tree G is a Steiner vertex. In view of (4) it is enough to show that $d(S_j)$ cannot be 1 or 2. If $d(S_j) = 1$, as in the top part of Fig. 8.10d, we merely erase S_j and its edge. If $d(S_j) = 2$ and the two edges of S_j make an angle $<180°$, as in the middle of Fig. 8.10d, we erase S_j and both its edges and we reconnect by adding the dotted edge. If $d(S_j) = 2$ and the two edges of S_j are in a straight line, as in the bottom part of Fig. 8.10d, then S_j serves no useful purpose and so we just remove it while keeping the segment uv as an edge of G.

We are now in the position to prove the essential upper bound on the number k of additional vertices $S_1, S_2, \ldots, S_k$ of the minimal tree G. First, a census is made of the n initial given vertices $A_1, A_2, \ldots, A_n$: Let n_1 have degree 1, n_2 degree 2, and n_3 degree 3. By (4) this completes the total count so that

$$n_1 + n_2 + n_3 = n. \tag{5}$$

The total number N of edges of G is counted twice but in two *different* ways. Since G is a tree with $n + k$ vertices in all, it follows from (3) that

$$N = n + k - 1. \tag{6}$$

Next, the minimal tree G has

$$n_1 \quad \text{vertices of degree 1,}$$

$$n_2 \quad \text{vertices of degree 2,}$$

$$k + n_3 \quad \text{vertices of degree 3.}$$

Multiply each number of vertices by the degree, to form the expression

$$n_1 + 2n_2 + 3(k + n_3).$$

What does this count? The answer is that it counts the total number N of edges but each edge gets counted twice, once from each end. Hence

$$n_1 + 2n_2 + 3(k + n_3) = 2N$$

or, in view of (5),

$$n + n_2 + 2n_3 + 3k = 2N. \tag{7}$$

Now it follows from (6) and (7) that

$$n + n_2 + 2n_3 + 3k = 2(n + k - 1),$$

so that

$$k = n - 2 - n_2 - 2n_3.$$ (8)

In particular, since n_2 and n_3 cannot be negative,

$$k \leq n - 2.$$ (9)

As we shall show, the importance of this bounding relation (9) is that it will enable us to perform an essential reduction of our problem. Roughly speaking, this reduction replaces one problem for n points known and k points unknown by a large but finite number of problems with all points known. That is to say, we shall solve our problem by examining a large number of candidates for the minimal tree G but each candidate will be a tree *all* of whose vertices will be known.

Before tackling the general case of n points $A_1, A_2, \ldots, A_n$ we examine what happens for certain small values of n. The case $n = 3$ is already known: It is Problem 6 of Chapter 6. So let $n = 4$; from (9) it follows that here

$$k = 0, 1, \text{ or } 2.$$

In fact, all three values are possible, as is shown in Fig. 8.11. By means of equation (8) the point-count balances: $n = 4$ and for

	k	n_2	n_3
Fig. 8.11*a*	0	2	0
Fig. 8.11*b*	0	0	1
Fig. 8.11*c*	1	1	0
Fig. 8.11*d*	2	0	0

so that (8) holds in every case. These four cases comprise all the possibilities for $n = 4$. There is clearly no difficulty in constructing the minimal

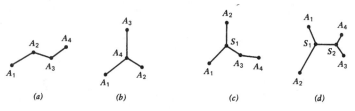

(a)   *(b)*   *(c)*   *(d)*

Figure 8.11

tree T for Fig. 8.11a and b. From our previous work we also know how to construct the net for Fig. 8.11c: by means of the basic equilateral construction shown in Fig. 6.7b of Chapter 6. But how do we find the minimal tree for Fig. 8.11d? Recall that only the points A_1, A_2, A_3, A_4 are known. How are S_1 and S_2 to be found?

When this is examined in detail, there appears the diagram of Fig. 8.12a. The basic equilateral construction determines the points A_{12} and A_{34}, which are the third vertices of their equilateral triangles. Then the

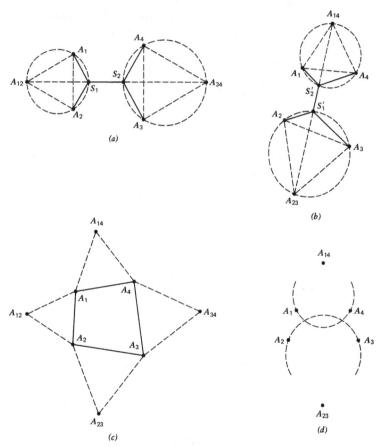

Figure 8.12

straight segment $A_{12}A_{34}$ is drawn and the two additional Steiner vertices S_1 and S_2 are the points in which the circumcircles of the two equilateral triangles cut $A_{12}A_{34}$. But we may also vary the construction to what is shown in Fig. 8.12b: A_1 is paired with A_4, and A_2 with A_3. This results in another candidate for the minimal tree, with the two additional vertices S_1' and S_2' lying on $A_{14}A_{23}$. Which of the two trees is shorter?

As a consequence of Ptolemy's theorem, it was proved in Chapter 1 that

$$A_1S_1 + A_2S_1 = A_{12}S_1 \quad \text{and} \quad A_3S_2 + A_4S_2 = A_{34}S_2;$$

thus the total length of this whole tree is

$$A_1S_1 + A_2S_1 + S_1S_2 + A_3S_2 + A_4S_2 = A_{12}A_{34}.$$

Similarly, the total length of the tree of Fig. 8.12b is $A_{14}A_{23}$. This leads to the following recipe: Given the four points A_1, A_2, A_3, A_4 as in Fig. 8.12c, build an outward equilateral triangle on each side of the quadrilateral $A_1A_2A_3A_4$, to form the quadrilateral $A_{12}A_{23}A_{34}A_{14}$. Find which of its two diagonals is shorter, $A_{12}A_{34}$ or $A_{23}A_{14}$. This shorter length is then the length of the minimal tree T and the two additional vertices lie on that shorter diagonal. If both diagonals are of equal length, then there are the two different minimal trees. Of course, it might happen that only one tree of such type exists; this is illustrated in Fig. 8.12d. Here the circumcircles of the equilateral triangles $A_1A_4A_{14}$, $A_2A_3A_{23}$ intersect; therefore we *cannot* pair A_1 with A_4 and A_2 with A_3.

Thus our solution of the problem for $n = 4$ depends, essentially, on choosing between two possible candidates. As will be seen later, the solution for general n will also depend on choosing between a number of alternative candidates. Roughly put, this choice comes from the fact that for our n points A_1, A_2, . . . , A_n we do not know in advance their interconnection scheme: We do not know which ones are to be paired in the minimal tree.

Consider next the minimal tree for the case of 10 points A_1, A_2, . . . , A_{10} in Fig. 8.13a. Here no vertex A_i has degree 3 so that $n_3 = 0$, and $n_2 = 4$ because there are four A_i-vertices of degree 2: A_2, A_5, A_6, A_8. Hence by (8) $k = 10 - 2 - 4 = 4$; the four additional vertices S_1, S_2, S_3, S_4 are shown. The 10 original points A_i split into three groups: the group A_1, A_2, A_3, A_4, A_5; the group A_6, A_7, A_8; and the residual group A_9, A_{10}. When each of the first two groups is considered *by itself*, it has the maximum possible number of S-points: $5 - 2 = 3$ for the first group,

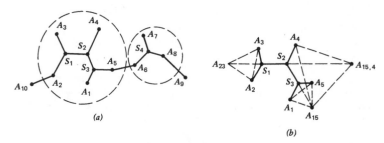

Figure 8.13

$3 - 2 = 1$ for the second. We call any such group of A_i points, connected together by a part of the minimal tree with its full quota of S-points, an *S-component*. Thus in Fig. 8.13a there is one S-component of five points, another S-component of three points, and the residual set of two points.

There is also an alternative description of S-components in terms of rigidity, or stability, which we mention very briefly. Suppose that the point A_4 of Fig. 8.13a is displaced in an arbitrary direction to a new position A_4' very close to A_4. How much of the minimal tree is changed as a result of that perturbation? Not all of it, but only the S-component containing A_4. So the S-components might also be called the variability sets.

The last special case considered is that of five points A_1, A_2, A_3, A_4, A_5 of Fig. 8.13b. This is just the large S-component of Fig. 8.13a, but considered all by itself. The construction of the minimal tree here results simply from applying the basic equilateral construction three times. Of course, we have to know *how* to associate the A_i points:

$$(A_2A_3)((A_1A_5)A_4). \qquad (10)$$

These bracketings produce successively the points A_{23}, A_{15}, and $A_{15,4}$, and now the total length of the minimal tree is $A_{23}A_{15,4}$. The same way works for any single S-component: Its minimal tree is obtained by performing the basic equilateral construction successively according to the correct association schema of the type (10).

We turn finally to the solution of our problem for the general case of n points. One thing must be made clear in advance: Our way of solving the problem uses only the elementary ruler-and-compasses geometrical constructions in the traditional sense, but it may use very many of them if n is at all large. Briefly put, we shall produce not a *practicable* geometrical algorithm but only a *finite* geometrical algorithm. That is to say, we cannot guarantee that our problem can be solved for a reasonably large value of n, say 20 or 30, in a reasonable time, even with the help of the largest and fastest computing machines in existence. All that we can be sure of is that the solution will be produced in a finite number of steps.

The solution algorithm consists of three parts: splitting the n points A_1, A_2, . . . , A_n into S-components and the residual set, constructing the part of the minimal tree over each S-component separately, and finally connecting together the resulting parts of the minimal tree and the points of the residual set.

Suppose then that the n given points A_i split up into S-components C_1, C_2, . . . , C_p and the residual set C_0. We do not know how to perform this splitting in advance. However, some simple rules can be shown to apply. For instance, no point A_i can belong to more than three different components (since its degree is ≤ 3), every component contains at least three different A_i points, and so on. These two rules alone show that the number of different splittings of A_1, A_2, . . . , A_n into S-components is finite. Therefore a list of all of them can be made and examined one by one.

Next, let us consider any particular splitting, and in this splitting let us single out a particular S-component C. We take up now the matter of constructing the minimal tree for C. As was shown earlier, this partial minimal tree can be constructed, provided that we know its association schema, of the type (10). Again, this association is not known in advance, but it is a simple matter to show that there is only a finite number of different associations for the A_i points belonging to C. So a list of all possible associations is made out. We examine these candidates one by one, and choose the association that produces the shortest minimal tree for C. Of course, not every candidate need produce a connecting network of the required type: a counterexample for four points A_1, A_2, A_3, A_4 is shown in Fig. 8.12d, where the association

$$(A_1A_4)(A_2A_3)$$

fails to produce our type of connecting net. It might even happen that there is no correct type of net for *any* association. This simply means that the particular candidate C cannot be an S-component. Thus, any splitting in which C occurs may be disregarded.

Finally, the same procedure is used for the last part: interconnecting the partial minimal trees over the *S*-components and the points of the residual set. Here too the number of possibilities is finite. Thus the minimal tree for $A_1, A_2, \ldots, A_n$ (or all the minimal trees, if there are several) will eventually be produced.

On examining our problem we see that it has a *continuous* ingredient (the $2k$ coordinates of the additional vertices S_j) and a *discrete* ingredient (interconnections—which points are connected to which). What our procedure really accomplishes is to remove that *continuous* ingredient. Unfortunately, this is done at the cost of greatly complicating the discrete ingredient.

EXERCISES

1. Show that every graph on four vertices can be drawn in the plane without crossed edges. Is the same true for five vertices?

2. Show that the relation (3) does not characterize trees among all graphs or among all special graphs. Does it characterize trees among all cycle-free graphs?

3. Show that if a graph *G* has an Euler circuit, then all of *G* can be split into a number of cycles of *G* so that no two cycles share an edge (though they may share vertices). Formulate and prove the converse.

4. A (special) graph *G* is called bipartite if its vertices belong to two (nonempty) groups such that two vertices in *G* are connected by an edge if and only if they belong to different groups. Show that a bipartite graph is triangle-free. Show that for every positive integer *k* there is a triangle-free graph with $2k$ vertices and k^2 edges.

5. Generalize the preceding problem to *p*-partite graphs. (*Hint*: Triangles are complete 3-graphs; now use complete *p*-graphs in place of triangles.)

6. For the seven Königsberg bridges of Fig. 8.2*a* and *b* can one bridge be (a) added, (b) removed, (c) replaced, so that an Euler path is possible? So that an Euler circuit is possible?

7. The graph *G* consists of the 8 vertices and the 12 edges of a cube. Show that *G* has 4-cycles, 6-cycles, and 8-cycles. Are every two *k*-cycles congruent? Does *G* have a Hamiltonian circuit?

8. The graph *H* consists of the six vertices and the 12 edges of a regular octahedron. Find an Euler circuit for *H*.

9. Can graphs equivalent to *G* and *H* above be drawn in the plane so that all edges are straight and no two cross?

10. Find the length of the shortest network that connects the four vertices of a unit square. How many such minimal networks are there?

11. Describe the shortest network connecting the five vertices of a regular pentagon.

12. Describe the shortest network connecting the four vertices of a regular tetrahedron.

13. For the Steiner problem give examples where one A_i point belongs to (a) two different *S*-components, (b) three different *S*-components. In view of such examples how would you describe *S*-components as variability sets?

CHAPTER NINE
ELEMENTS OF CONVEXITY

This chapter is concerned with an informal description of convexity and certain related concepts. The aim is to provide some intuitive understanding and to motivate the definitions by means of some examples and counterexamples in two and three dimensions.

It may be claimed that geometry is concerned with (an idealization of) the *visible* world, and the basic informal definition of a convex figure F illustrates that claim:

(D_0)   F is convex if and only if any two points of F are mutually visible via F.

For instance, an ordinary rectangular room is convex (if empty) but an L-shaped room is not. In the plane the interior of a circle is convex but a ring-shaped domain consisting of all points between two concentric circles is not. The inside of a triangle is always convex but the inside of an *n*-gon need not be convex if $n > 3$. The intuitive description

$$\text{convexity of } F = \text{unobstructed visibility via } F$$

is simply formalized to a definition. If a and b are points, let $[ab]$ denote the straight segment from a to b, endpoints included. Then the informal definition (D_0) converts to the following formal definition

(D_1)   *F* is convex if and only if the whole of $[ab]$ is in *F* whenever a and b are in *F*.

Normal behavior, whether of people, animals, things, or concepts, is illustrated by their abnormal behavior; thus a better idea of what convexity is may be obtained by having a brief look at what it is not. So we shall briefly consider some simple and conspicuous ways in which figures

175

fail to be convex. Clearly, a set consisting of two, or more, separate pieces cannot be convex because two points lying in different pieces are not mutually visible via the set. This could be formalized to the statement that a convex set must be connected (= all of one piece). Consider next the two examples

$$X = \{(x, y): 0 < x^2 + y^2 \leq 1\},$$

$$Y = \left\{(x, y): x^2 + y^2 \leq 1, (x - \tfrac{1}{2})^2 + y^2 \geq \frac{1}{100}, x^2 + (y + \tfrac{1}{2})^2 \geq \frac{1}{225}\right\}.$$

X consists of all points inside or on the unit circle C about the origin, except for the origin itself. Y consists of all points inside or on C but with the points inside certain two smaller circles in C removed. Both X and Y fail to be convex because the removed parts of C prevent certain pairs of points from being mutually visible. More generally, it can be said that X and Y fail to be convex because, even though they are connected, they are not simply connected. That is, one can make a closed tour inside X or Y, which goes around certain points not belonging to the set. Equivalently, not every closed curve in X or in Y can be shrunk *inside its set* to a point. Or, equivalently again, a pair of points in X or in Y can be joined by two different arcs in the set so that neither arc can be continuously moved *inside its set* onto the other. Recall that similar behavior was observed earlier with the plane sections of a torus. It is clear that if that sort of thing happens for a set, then some pairs of its points are mutually obstructed by a foreign "island." We obtain thus the beginnings of a suggestive descending hierarchy:

sets → connected sets → simply connected sets

→ linearly connected, or convex, sets.

This hierarchy may be extended by using the further classes of arcwise connected sets and of polygonally connected sets. It can be shown that a set may be connected without being arcwise connected; the standard example is

$$\left\{(x, y): y = \sin\frac{1}{x}, x > 0\right\} \cup \{(0, y): -1 \leq y \leq 1\}.$$

This consists of two parts; the first part is a sinusoidal-type curve oscillating between lines $y = 1$ and $y = -1$ faster and faster as x approaches 0; the second part is the vertical segment $-1 \leq y \leq 1$ on the Y-axis. Each part, considered separately, is connected (and even arcwise connected); the whole is also connected because any point of the second part is a limit point of a sequence of points in the first part. However, the whole set is not arcwise connected because no arc belonging to the set can join a point in one part to a point in the other part. Similarly, there are sets that are arcwise connected without being polygonally connected; an example is

$$\{(x, y): 0 \leq x \leq 1, x^3 \leq y \leq x^2\}.$$

Here the point (0, 0) cannot be polygonally joined to any other point of the set.

It so happens that all the properties used to describe the various classes of the preceding sets are of the following single generic type:

The set X is such that, together with any two points, it also contains a certain special subset of X joining those two points.

The following list can now be made:

Property of X	The joining set is
Connectedness	"One piece" (= connected)
Arcwise connectedness	An arc
Simple connectedness	An arc joing the two points, and deformable inside X to any other such arc
Polygonal connectedness	A piecewise linear arc
Convexity	A linear arc

This exhibits convexity as a very special type of connectedness obtained by an extreme specialization of the joining set: It must be a straight segment.

Unless the contrary is explicitly stated, we shall assume from now on that the sets considered are always *closed*. This means that they include their boundaries or, equivalently, that the limit point of any convergent sequence of points belonging to the set belongs itself to the set. Alternatively, it is possible to define first an *open* set as one that, together with any point, also contains all points sufficiently close to that point. Then a set is defined to be closed if and only if its complement is open. Why make the assumption that all our sets are closed?

First, nothing is lost or gained in the way of metric properties, which are always of considerable interest. It must be emphasized that this is so only for convex sets, and not for general sets. For instance, let X be the set of all rational numbers between 0 and 1. This X, considered as a subset of the real line, is neither open nor closed; being countable, it has linear content, or measure, 0. The closure of X is the closed interval [0, 1] whose measure is 1. On the other hand, the three intervals

$$0 < x < 1, \qquad 0 \leq x < 1, \qquad 0 < x \leq 1,$$

all of which are convex sets of the line, have the same length 1 as the closed interval $0 \leq x \leq 1$. To put it roughly, we might say that the process of adjoining its boundary to a convex set X does not change its size, content, surface, and so on (though it may do so if X is not convex).

Second, certain alternative definitions of convexity by means of separation or proximity, which we shall provide shortly, are considerably simpler for closed convex sets.

Third, the assumption of being closed excludes such unintuitive examples as

$$A = \{(x, y): x^2 + y^2 < 1\} \cup X,$$

which consists of all points inside the unit circle C, plus an *arbitrary* subset X of C. The set A satisfies our definition (D_1) of convexity; observe that, no matter what X is, A has the same "circumference" and "area" as the closed set $\{(x, y): x^2 + y^2 \leq 1\}$.

As an example of dealing with closed sets only, we have the following alternative definition of convexity:

(D_2) A set X is convex if and only if it contains the midpoint of any two of its points.

For if a and b are two points in X, then the following points are also in X: the midpoint m of $[ab]$, the two midpoints of $[am]$ and $[mb]$, the next four midpoints, and so on. In this way one gets a finite but arbitrarily dense collection of successive midpoints on $[ab]$. Since X is closed by hypothesis, it follows that the whole segment $[ab]$ is in X; therefore X is convex by (D_1). Conversely, if (D_1) holds, then obviously (D_2) also holds.

As was mentioned before, not every plane polygon P_n is convex if $n > 3$. Here by the polygon P_n we mean really the polygonal domain: all the points on the periphery as well as inside. Thus every P_n is closed, since it contains its boundary. A general polygon P_n is convex if and only if all its (interior) vertex angles are $< \pi$; this is automatically true if n

= 3, that is, for triangles. We can also express it differently: A polygon P_n is convex if and only if it is not reentrant. The concept of reentrant parts extends easily to plane domains more general than polygons: beside the reentrant quadrilateral of Fig. 9.1a we have also reentrant domains of Fig. 9.1b or c. This leads us to the following informal description of plane convex sets: They are

connected, simply connected, nonreentrant.

Next the phenomenon of reentrance will be characterized more formally, and this will lead to another alternative definition of convexity. For the convex domain X of Fig. 9.1d we observe that every point outside X can be strictly separated from X by a straight line, as is shown in the figure

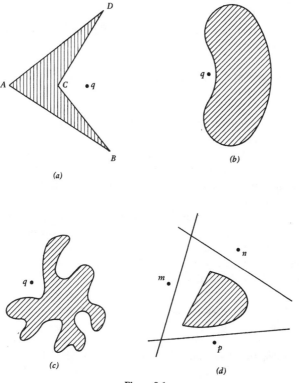

(a)

(b)

(c)

(d)

Figure 9.1

for the points m, n, p. The domains of Figs. 9.1a, b, c are connected and simply connected, yet for each of them there are points q, as shown in the figures, that are external but cannot be separated from the domains by straight lines. Hence the definition: A plane set X is reentrant if and only if there are points outside X that cannot be separated from X by a straight line. Our restriction to closed sets is handy here: The interior of a circle is an *open* convex set that cannot be separated by a straight line from any point on the circle. It can be shown that a set which is not connected, or not simply connected, must be reentrant according to our definition. In fact, we have the following new definition of (plane closed) convex sets by means of rectilinear separability:

(D_3) X is convex if and only if every point outside X is separable from X by a straight line.

For sets in three dimensions we merely replace the straight lines by planes, and then

(D_4) X is convex if and only if every point outside X is separable from X by a plane.

There is a different, and perhaps unexpected, way of excluding reentrant parts, which leads to another definition of convexity. Consider any nonconvex set X such as, for instance, those of Fig. 9.1a, b, c, and let a circle K of sufficiently big radius roll on the outside of X. Then there must be positions of K when it is in contact with more than one point of the boundary of X. Let c be the center of K when it is in such multiple-contact position. This means that there are in X at least two different points that realize the minimum distance from c. On the other hand, any circle, no matter how large or small its radius, can be rolled on the outside of a convex set so that it has at all times a one-point contact. Hence comes an alternative definition:

(D_5)   A set X is convex if and only if any point outside it has a *unique* point of X closest to it.

We observe that this holds for the plane as well as three-dimensional space—all that is needed is to replace the circle K outside X by a sphere.

Further, in (D_5) we also use our restriction to closed sets X; otherwise the minimum distance might not even be attained.

Let us determine the possible types of convex sets of a straight line. This is very easy to do since for a linear set all the connectivity properties we have considered coincide: connectedness, simple connectedness, arc-wise connectedness, polygonal connectedness, and convexity are *here* one and the same thing. Without the restriction to closed sets there are 11 types of convex linear sets:

$$[a], \quad (ab), \quad [ab), \quad (ab], \quad [ab],$$
$$(-\infty, a), \quad (-\infty, a], \quad (b, \infty), \quad [b, \infty), \quad (-\infty, \infty), \quad \emptyset \quad (1)$$

that is, a point, an open interval, two semiopen intervals, a closed interval, four types of semi-infinite intervals or rays, and the whole line. The last case is the empty set $\emptyset$, added to the list for technical convenience, as will be seen shortly. If we limit ourselves to closed sets, then there are only six types of convex linear sets:

$$[a], \quad [ab], \quad (-\infty, a], \quad [b, \infty), \quad (-\infty, \infty), \quad \emptyset. \quad (2)$$

Here again the empty set is included for technical convenience. The reason for our brief excursion into convex linear sets is that they lead to a convenient new definition of convexity:

(D_6) A set X is convex if and only if any straight line L intersects it in a convex linear set.

This definition holds in the plane as well as in three-dimensional space. Also, for general convex sets the convex linear set must be one of the list (1), whereas for closed convex sets it is one in (2). We see now the advantage of having the empty set included in both (1) and (2); this takes care of the case when L does not intersect X at all.

Next comes the important concept of the (closed) convex hull $\bar{X}$ of an arbitrary set X; here X itself need not be closed or connected. First let X be a plane set. The idea behind forming $\bar{X}$ out of X is in line with our informal description of convexity as unobstructed visibility: Since certain pairs of points of X are not mutually visible via X, we get the convex hull $\bar{X}$ by adjoining to X all obstructing points. Or $\bar{X}$ consists of all points inside or on a thin elastic rubber band stretched around X. More formally, three equivalent definitions of $\bar{X}$ can be given:

(A) $\bar{X}$ is the smallest convex set containing X.
(B) $\bar{X}$ is the part common to all half-planes containing X.
(C) $\bar{X}$ is the closure of the region $C(C(X))$, where $C(U)$ stands for all the points of straight segments with endpoints in the set U.

In definition (A) it must be explained what "the smallest convex set" means, and the explanation hinges on one of the most important features of convexity: It is preserved under arbitrary intersections. That is, the part common to any collection of convex sets is itself convex. For instance, consider all parallel infinite strips of width 2 in the plane, such that the origin lies halfway between the two bounding lines; a few of such strips are shown in Fig. 9.2a. Each strip is a closed convex set; the intersection of all of them, that is, the part common to all, is the unit circle together with its inside, and it is also a closed convex set. Our assertion that intersections preserve convexity is easily proved: let a and b be two points, each of which lies in every convex set of the collection, then $[ab]$ also lies in each set and hence it belongs to the common part of them all, hence the latter is convex.

   In view of our restriction to closed sets only, it is important to note that the same preservation under intersections holds here: The part com-

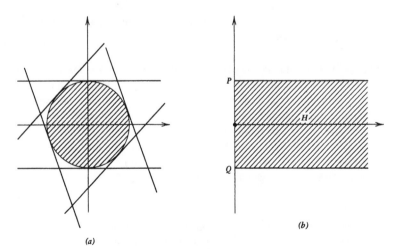

(a)

(b)

Figure 9.2

mon to any collection of closed sets is itself closed. In technical language the preservation property is expressed as follows: Let $\mathscr{C}$ denote the family of all closed convex sets,

$$\text{if } X_\alpha \in \mathscr{C} \text{ for every } \alpha, \text{ then } \cap X_\alpha \in \mathscr{C}.$$

Or, even more briefly,

$$(\alpha)X_\alpha \in \mathscr{C} \rightarrow \cap X_\alpha \in \mathscr{C}.$$

This is read: If, for every α, X_α is closed convex, then the intersection of all X_α's is also closed convex. It is now simple to define the smallest convex set containing X: the intersection of *all* (closed) convex sets containing X. For, such intersection is closed, convex, containing X_α, and by its very definition it is the smallest such set, being contained in any other set that is closed, convex, and contains X. In symbols, definition (A) is

$$\bar{X} = \bigcap_{\substack{Y \in \mathscr{C} \\ X \subseteq Y}} Y.$$

For definition (B) a half-plane means a closed half-plane: the set of all points in the plane that lie on a straight line or to one side of it; this concept was used before in the discussion of densest packings and Voronoi regions. Definition (A) defines $\bar{X}$ as the intersection of *all* closed convex sets containing X, whereas (B) defines it as the intersection of only certain *special* such sets, namely, closed half-planes. It follows that $\bar{X}$ as defined by (B) *contains* $\bar{X}$ as defined by (A) because the intersection of *fewer* factors is never smaller than the intersection of *more* of them. However, the reverse also holds: $\bar{X}$ as defined by (A) contains $\bar{X}$ as defined by (B). For otherwise there would exist a point q belonging to every closed half-plane that contains X, but separable from X by a straight line, which leads to a contradiction. Thus the definitions (A) and (B) are equivalent.

For the definition (C) we show first why the collecting operation C has to be performed twice. Let X be the set consisting of the three vertices of a triangle. Then $C(X)$ consists exactly of the three sides of the triangle but without its interior. $C(C(X))$ gives us the whole closed convex hull: the triangle with its interior. Next, we show why $\bar{X}$ is defined as the closure of $C(C(X))$, i.e., why the boundary of $C(C(X))$ must be adjoined. In our previous example of the triangle this is not necessary. However, consider the example of Fig. 9.2*b*, where X consists of the two points

$P(0, 1)$ and $Q(0, -1)$, and of the whole semiaxis $H = \{(x, 0): x \geq 0\}$. Here the closed convex hull $\bar{X}$ is the horizontal semifinite strip but $C(C(X))$ consists only of the interior of that strip together with the vertical segment PQ. Thus the horizontal rays from P and from Q have to be adjoined to yield a closed set. Definition (C) can be shown to be equivalent to (A) and (B). The definition (A) uses convexity but (B) and (C) do not. It is therefore possible to reverse the order of procedure: Define first convex hulls by (B) or (C), and then define a set to be convex if and only if it happens to coincide with its own convex hull.

For sets X in three dimensions the convex hulls $\bar{X}$ are defined as before, with very slight alterations. The definition (A) goes through unchanged. In (B) the half-planes are merely replaced by half-spaces; these are closed regions consisting of all points on, or to one side of, a plane in space. Definition (C) also turns out to apply in three dimensions without any change. To show this, one may use a theorem due to C. Caratheodory. It follows from this theorem that if X is any set in the ordinary three-dimensional space, then any point p of the convex hull $\bar{X}$ of X belongs to a tetrahedron T whose vertices are in X. Next it is shown that if the vertices of T belong to X then the whole tetrahedron T is a subset of $C(C(X))$. Therefore the definition (C) is valid as it stands.

As was mentioned before, the convex hull $\bar{X}$ of a plane set X may be conveniently visualized as the region of points lying on or inside a thin elastic rubber band stretched around X. The following illustration may help with visualizing the convex hull $\bar{X}$ of a set X in three dimensions. Imagine X to be a very hard core completely imbedded in some solid but soft material, so as to form a lump S. We have at our disposal a very large flat grinding wheel and we use it so as to grind down as much of S as possible. The grinding wheel is softer than the core X and cannot remove any part of it, but it is harder than S. What remains after everything possible has been ground off S? Precisely the convex hull $\bar{X}$, in virtue of definition (B).

We take up next the following problem: What closed convex sets X in three dimensions contain a complete straight line A? Let S be the section of X by any plane Q at right angles to A. Since both X and Q are closed and convex, so is their intersection S. Next, let p be any point in S and let A_p be the straight line through p, parallel to A. Since p and A belong to the convex set S, it follows from the definition of convexity that X contains every straight segment $[pa]$, where a is any point of A. The totality of such segments $[pa]$ contains the open plane infinite strip bounded by the two lines A and A_p. But X is also closed; therefore it must include the boundary of that strip; it follows that X contains A_p. Thus X consists of points on straight lines, parallel to A, and crossing S. Hence X is a solid cylinder with the base S, in the direction of A.

It may occur to us to ask a related question, about closed convex sets in three dimensions, which contain *two* complete straight lines, say L and M. To make it a "good" problem, the lines L and M should neither intersect nor be parallel: They are skew. For definiteness we formulate our problem as follows: to find the closed convex hull $\bar{X}$ of the set X consisting of L and M. This is done very simply by reference to the previous problem. Let p be any point in $\bar{X}$; then, by the preceding, $\bar{X}$ contains the line L_p through p that is parallel to L, as well as the line M_p through p that is parallel to M. Therefore, being convex, $\bar{X}$ contains the whole plane through p that is parallel to L and M simultaneously.

Hence the answer is as follows. Let U be the plane containing L and parallel to M, let V be the plane that contains M and is parallel to L. Then $\bar{X}$ is the closed infinite slab consisting of U, V, and all points between U and V. Our problem could be changed into the following form: What is the set of points of all straight closed segments whose endpoints lie on L and M? In terms of the notation used before, this set is $C(X)$. The answer is now not quite the same as before: $C(X)$ is the set of all points strictly between U and V, together with L and M. Thus $C(X)$ is neither open nor closed.

The process of passing from a set X to its convex hull $\bar{X}$ is an augmenting process: Unless X itself is convex already, something must be added to X to form $\bar{X}$. We consider now a subtractive process that is nevertheless related to the preceding. We ask, "Given an arbitrary set X, how much can be taken away from it so that the remainder still has the same convex hull as X itself?" This will lead us to the important notion of extreme points of X.

The best way to approach our reduction process is by means of the collecting operation $C(U)$, where $C(U)$ consists of all straight segments whose endpoints are in U. Now reverse this collecting operation: Whenever X contains a straight segment, remove it *except* for its endpoints. What remains is precisely the irreducible core of X we want, the smallest subset of X that has the same convex hull as X. This leads to the definition: A point of X is called extreme if and only if it is not an interior point of a straight segment belonging to X. What we have informally called the irreducible core of X is just the set of extreme points of X. For instance, in the plane the extreme points of any convex polygon (whether with or without its interior) are its vertices. The same is true for a convex polyhedron in space. In three dimensions, the extreme points of a solid ball B form the sphere S that is the boundary of B. If we take an ordinary right circular cone, with a circular base, and regard it either as a solid or as a surface bounding that solid, then the extreme points consist of the vertex and of all points on the circular rim of the base. Note, however, that we cannot define the extreme points of a complete, i.e., infinite, cone. In general, extreme points are defined only for *bounded* sets.

A certain technical difficulty arises in connection with extreme points: The set of all extreme points of X need not be closed even though X is closed or both closed and convex. To show this on an example, we shall first introduce a certain solid that lies, so to speak, halfway between an ordinary circular cylinder and an ordinary circular cone. All three solids have circular bases; a cylinder has a circular top, a cone has a point top, and the new solid has a straight segment as top. For proper definition consider an ordinary circular cylinder whose base is a circle C and whose top is a (congruent) circle C_1. Let v be the center of C_1 and let D be any diameter of C_1. Then the convex hulls of the three sets

$$C \cup C_1, \quad C \cup \{v\}, \quad C \cup D$$

are, respectively, a cylinder, a cone, and an example of the new solid. By the reason of its shape the latter will be called simply (though not accurately) a chisel. It is clear what is meant by the height or the base radius of any one of our three solids. Although the lateral surfaces of all three are made up of straight segments, an important difference may be observed. A cylinder or a cone can be rolled on a plane and so their surfaces can be *developed*, i.e., unrolled onto a plane exactly, without distortion. A chisel cannot be rolled on a plane, and its surface is not developable. The usual example of a surface made up of straight lines but not developable, is the one-sheeted hyperboloid. A special instance is a one-sheeted hyperboloid of revolution. This surface arises when we take two skew straight lines L and M, and rotate M about L as axis. However, the example of the surface of a chisel shows that a nondevelopable surface made up of straight lines can even be the boundary of a convex region.

We return now to our counterexample for extreme points. Take two chisels of the same base radius and the same height, and join them base to base, let the resulting solid be called X. What is the set E of its extreme points? Obviously E includes the four points v_1, v_2, v_3, v_4, which are the endpoints of the two edges; suppose that $v_1 v_2$ is the upper edge, that $v_3 v_4$ is the lower edge, and that v_1 lies above v_3 and v_2 lies above v_4. What other extreme points are there? They must all lie on the circle C, which is the rim of the common base of the two chisels making up X. However, *not all* points of C are extreme: There are exactly two exceptions, the midpoint of $v_1 v_3$ and the midpoint of $v_2 v_4$. Thus the set E of extreme points of X is not closed even though X itself is closed and convex.

We introduce now our last two concepts related to convexity: star-shaped sets and the convex hub $\check{X}$ of a set X. First, we go back to the very beginning, and we recall that a set X is convex if and only if any point of it is visible from any other point, unobstructed. The concept of a star-shaped set arises when the requirement of unobstructed visibility is changed to a less restricted form: A set X is *star-shaped* at its point p if and only if every point of X is visible from p via X. Another name for this also exists: We say that X is locally convex at p. For instance, the quadrilateral of Fig. 9.3a is star-shaped at every point of the smaller cross-hatched quadrilateral. Similarly, the set bounded by the smooth (but reentrant) closed curve C in Fig. 9.3b is star-shaped at every point of the cross-hatched quadrilateral. Here it may be observed that the four straight lines that carry the sides of the quadrilateral are the inflexion tangents of the boundary curve C. A set can be star-shaped at all points of a segment only, as in Fig. 9.3c, or it can be star-shaped at a single point, as in Fig. 9.3d. Finally, a set X can fail to be star-shaped at any point, as is shown in Fig. 9.3e. The set of all points in X at which X is star-shaped is called the convex hub $\check{X}$ of X. As the examples of Fig. 9.3 suggest, $\check{X}$ itself is always convex. This is easily proved, for if $p \in \check{X}$ and $q \in \check{X}$ then the whole straight segment $[pq]$ must belong to $\check{X}$. From our definitions it follows that any set X contains its convex hub and is contained in its convex hull:

$$\check{X} \subseteq X \subseteq \bar{X}.$$

If on either side there is an equality, that is, if

$$\check{X} = X \quad \text{or} \quad X = \bar{X},$$

then all three sets coincide and X is convex itself.

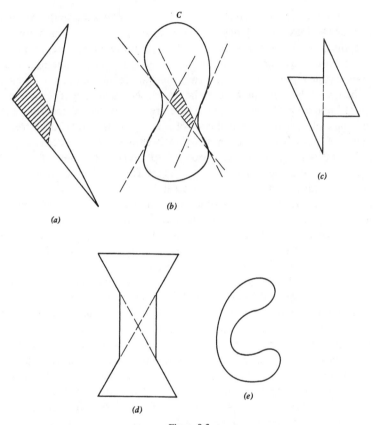

Figure 9.3

EXERCISES

1. Let C be a convex polygon and let $C(a)$ be the plane region covered by all circular disks of radius a with center on or in C. Show that

$$\text{area } C(a) = \text{area } C + a \text{ length } C + \pi a^2.$$

2. Let Z be a circular cylinder of base radius r and height h, and let C, C_1 be the circular rims of its base and its top. Find the maximum volume of a tetrahedron T that can be inscribed into Z. (*Hint:* First show that the vertices of T are extreme points of Z, then consider two cases.)

3. Let C be a chisel of base radius r and height h, let E be its edge and B its base. Show that the sections of C by planes perpendicular to E are isosceles triangles of height h and the sections of C by planes parallel to B are ellipses of major semiaxis r.

4. Using the preceding exercise, show in two different ways that the volume of C is $\pi r^2 h/2$.

5. Using the preceding two exercises, justify the following statement: The volumes of a cylinder, a chisel, and a cone, of base radius r and height h are

$$\pi r^2 h, \quad \frac{\pi r^2 h}{2}, \quad \frac{\pi r^2 h}{3}$$

because

$$\int_0^1 dx = 1, \quad \int_0^1 x \, dx = 1/2, \quad \int_0^1 x^2 \, dx = 1/3.$$

6. Let L and M be two skew straight lines in space. Show that the slab that is the closed convex hull of $L \cup M$ has the thickness d, where d is the minimum distance from a point on L to a point on M.

7. Let X be the set of the four vertices of a solid tetrahedron T. The set $C(X)$ consists of the six edges of T, and the set $C(C(C(X)))$ is all of T. Show that already $C(C(X))$ is all of T.

8. Describe the set of extreme points of a solid torus.

9. Let C_1 be a closed convex curve in the plane, lying entirely inside another closed convex curve C. Show that

$$\text{Length } C_1 \leq \text{length } C.$$

If no point of C_1 is closer than d to a point of C, show that

$$\text{Length } C_1 \leq \text{length } C - 2\pi d.$$

(*Hint:* C and C_1 are arbitrarily well approximable by inscribed polygons. Use exercise 1.)

10. Let S be a sphere and X a subset of S lying in an open hemisphere. How would you define spherical convexity of X? Is every spherical triangle (with its interior) spherically convex? Is every spherical quadrilateral spherically convex?

11. The convex hub $\check{X}$ of a plane n-gon X is empty. Show that $n \geq 6$.

12. If every two of the three straight lines L, M, N are skew and a set

X contains all three lines, show that the convex hull of X is the whole space.

13. Let a closed plane curve C have a polar equation $r = f(\theta)$, where f is a continuously differentiable function satisfying $0 < a \le f(\theta) \le b$, $f(2\pi) = f(0)$, $|f'(\theta)| \le K$. If X is the region bounded by C, show that the convex hub $\tilde{X}$ contains a circle.

14. Show that a chisel C cannot be rolled on a plane. (*Hint:* Observe what would happen to the edge of C during a hypothetical rolling.)

CHAPTER TEN

CURVES IN SPACE AND CURVES ON SURFACES

This section is devoted to some elements of classical differential geometry: a branch of geometry that uses calculus and linear algebra to investigate the behavior of curves and surfaces. Some knowledge of the rudiments of vector calculus and of linear algebra will therefore be assumed, and it will be seen how these two subjects interact to help us derive our principal result, the Frenet-Serret equations, which describe the curving and twisting of curves in space. Then we shall consider a few special cases of geodesics: curves of shortest length which join two points of a surface and lie entirely on that surface.

By way of introduction and motivation let us start with the following question: What property do the line L, the circle C, and the circular helix H have in common, that no other curve has? This question is rather loose and a variety of answers could be given. Or, perhaps, the one correct answer could be phrased in a variety of ways. Looking at the three curves locally, piece by piece, we observe that all three are somehow homogeneous or uniform, in the sense that a neighborhood of a point on any one curve looks exactly like the neighborhood of any other point on that curve. This self-congruence "in the small" suggests looking at the three curves "in the large" or globally, that is, looking at the whole curves. It might occur to us to move the whole curve; the motion is supposed to be rigid so that the curve may be imagined to be made of thin, stiff wire. We observe then that each curve acts as its own track, it can be moved "within" itself: L by being translated along itself, C by being rotated on itself, and H by a screw motion over itself. Or, to put the same thing differently, we can take two copies of any one of our three curves, one copy is stationary and we call it the track, the other copy is movable and we call it the train. It is now possible to move each train on its track without any distortion: the train and the track, as wholes, coincide at all times.

We propose to pursue the matter further, and for this purpose we introduce the concept of a local property of a curve. Let K be a general curve in space, given by a vector function $\bar{x} = \bar{x}(u)$, where u is some convenient parameter. For instance, if K is regarded as a trajectory of a point moving in space, then u could be the time t. Or u could be the arc length s measured on K from some fixed point. Then $\bar{x} = (x, y, z)$ is the vector of the Cartesian coordinates

$$x = x(u), \qquad y = y(u), \qquad z = z(u)$$

of the point on K corresponding to the value u of the parameter. It will be supposed that the vector function $\bar{x}(u)$ has as many derivatives as may be needed. Fix the value of u; this gives us a specific point p on K. We say then that a property of K at p is local if it depends only on the first few derivatives $\bar{x}'(u), \bar{x}''(u), \ldots, \bar{x}^{(k)}(u)$ at u. Recalling the definition of derivatives as limits, we see that a local property of K at p depends not on the whole of K but only on any neighborhood of p in K, that is, on an arbitrarily small arc of K that contains p as an interior point. This justifies the name "local property." The best-known local property of a curve K at a point p is of course its direction, that is, the direction in which K points at p. This is very well known and depends only on the first derivative $\bar{x}'(u)$. Of particular interest to us are two further local properties of curves in space: curving and twisting. Informally speaking, we describe curving as the tendency to depart from rectilinearity, and twisting as the tendency to depart from planarity. Later on, we shall give definitions of quantities that measure those properties: curvature, and torsion. It will turn out that curvature depends only on the first two derivatives $\bar{x}'(u)$ and $\bar{x}''(u)$ and torsion depends only on the first three derivatives $\bar{x}'(u)$, $\bar{x}''(u)$, and $\bar{x}'''(u)$.

Returning now to our three special curves L, C, and H, we shall briefly examine their self-congruence with regard to the local properties of curving and twisting. Since each curve can be made to ride over itself, so that any point of it can be moved by a self-congruence onto any other point, it must be true that all three curves have both constant curvature and constant torsion. The straight line L neither curves nor twists, so that its curvature and torsion are both zero; it therefore "lies on a line," in fact on itself. The circle C has constant nonzero curvature but zero torsion; hence it is a uniformly bent but plane curve, which can be rotated on itself. The helix H has constant nonzero curvature and constant nonzero torsion. So it is a curve that bends uniformly and twists uniformly; no piece of it, however small, lies in a plane. We note next that at any point L has the same direction while the directions for C and H vary from point to point. However, at any point of C the direction makes the same angle, $\pi/2$, with the radius vector that joins the center of C to the point in question. Also, at every point of H the direction makes the same angle with the axis of the circular cylinder on which H lies. Finally, a certain subordination of twisting to bending must be emphasized: a curve can bend without twisting, like the circle C or other plane curve, but it cannot twist without bending, for if it does not curve it is then a straight line.

We shall now set up the machinery toward proving the basic formulas for space curves, the Frenet-Serret equations, which describe the curving and twisting of space curves. They are named after two French mathematicians who have discovered them independently (F. Frenet in 1847, J. A. Serret in 1851). Originally, those equations applied to curves in the ordinary space of three dimensions. However, it turns out to be no harder, and perhaps even easier, to prove a generalization of Frenet-Serret equations, for a curve K in the n-dimensional Euclidean space E^n, and this is what we shall do. A curve K is accordingly given in its parametric form as $\bar{x} = \bar{x}(s)$ where $\bar{x}(s)$ is the vector of n rectangular coordinates

$$\bar{x}(s) = (x_1(s), x_2(s), \ldots, x_n(s)).$$

The parameter s is the arc length measured on K from some fixed point. It will be assumed that the first n derivatives

$$\bar{x}'(s), \bar{x}''(s), \ldots, \bar{x}^{(n)}(s) \tag{1}$$

exist. Here, as everywhere else, the marks $', ", \ldots, {}^{(n)}$ always denote differentiation with respect to s. Since the parameter s is the arc length, it follows that $\bar{x}'$ is a unit vector:

$$|\bar{x}'(s)| = 1.$$

The n derivatives (1) are n vectors which we imagine to originate at the point p on K corresponding to the parameter value s. It will be supposed that p is a *regular* point; this means that the n vectors (1) are linearly independent. Hence (1) is a basis, in the usual vector-space sense, of E^n.

Next we use the standard and very general technique of geometry and analysis: a more complicated structure is approximated locally by a simpler one of best possible fit. So, for instance, a curve is locally approximated by its line of best fit, that is, by a tangent line at a point of the curve. A surface is similarly approximated by a tangent plane at a point.

For better than linear fits we use the Taylor polynomials. In our case of a curve K in n-space E^n we introduce the successive osculating flats F_1, $F_2, \ldots, F_n$ at the point p of K. The first one, F_1, is one-dimensional and is spanned by the vector $\bar{x}'(s)$; it is therefore simply the tangent line to K at p. The next one, F_2, is two-dimensional and is spanned by $\bar{x}'(s)$ and $\bar{x}''(s)$, and so on. Generally, the k-dimensional flat F_k is set of all linear combinations

$$\sum_{i=1}^{k} a_i \bar{x}^{(i)}(s),$$

it is a k-dimensional plane in E^n passing through the point p at which we work. Each flat F_k is contained in the next one, F_{k+1}, and the last one, F_n, is the whole space E^n. The successive flats F_k may also be described as follows: F_k is the k-dimensional plane of best local fit to K at p. Or: F_k is the k-dimensional plane passing through $k + 1$ consecutive points of K at p. Now, as it stands, the last statement is complete nonsense because there is no such thing as "consecutive" points on K, let alone "consecutive" points at p. However, that statement is merely a convenient shorthand for the following: Consider $k + 1$ points on K with the indicated values of the parameter

$$p(s), p(s + \Delta s), p(s + 2\,\Delta s), \ldots, p(s + k\Delta s),$$

and let P_k be the k-dimensional plane determined by them; now let Δs approach zero so that all points tend to p; then F_k is the limiting position of P_k. Thus F_k is the best k-dimensional flat approximation to K at p.

It has been assumed that the n vectors (1) are linearly independent so that they form a basis in E^n. But we like to work with certain distinguished bases, the orthonormal ones. So we convert the basis (1) to an orthonormal basis

$$\bar{t}_1, \bar{t}_2, \ldots, \bar{t}_n \tag{2}$$

of n unit vectors which are pairwise orthogonal:

$$\bar{t}_i \cdot \bar{t}_i = 1, \qquad \bar{t}_i \cdot \bar{t}_j = 0, \qquad i, j = 1, 2, \ldots, n, \quad i \neq j. \tag{3}$$

The conversion is a standard and simple matter. Since s is the arc length, we have $|\bar{x}'(s)| = 1$ so that $\bar{x}'(s)$ is already a unit vector; we take $\bar{t}_1 = \bar{x}'(s)$. Now $\bar{t}_1$ lies in F_1 which is contained in the two-dimensional flat F_2, so we choose in F_2 a unit vector $\bar{t}_2$ perpendicular to F_1, then in F_3 we choose a unit vector $\bar{t}_3$ perpendicular to F_2, and so on. Eventually, we obtain the n vectors of the orthonormal basis (2). Like the vectors of

(1), the n vectors of (2) depend on s, i.e., they change from point to point on the curve K. We are interested in how the n vectors of (2) change with s, and the equations that describe this change are precisely the Frenet-Serret equations. Since (2) is a basis every vector is uniquely representable as a linear combination of the $\bar{t}_i$'s; in particular

$$\bar{t}_i' = \sum_{j=1}^{n} \rho_{ij}\bar{t}_j, \qquad i = 1, 2, \ldots, n. \tag{4}$$

It is the $n \times n$ matrix $R = (\rho_{ij})$ that is to be determined, and we determine its form quite simply, from the interplay of two basic pieces of information: one from linear algebra, the other from calculus. The algebraic piece of information is that R is skew-symmetric:

$$\rho_{ii} = 0, \qquad \rho_{ij} + \rho_{ji} = 0, \qquad i, j = 1, 2, \ldots, n. \tag{5}$$

Differentiating with respect to s the equations (3) we have

$$\bar{t}_i \cdot \bar{t}_i' = 0, \qquad \bar{t}_i' \cdot \bar{t}_j + \bar{t}_i \cdot \bar{t}_j' = 0$$

but by (4)

$$\bar{t}_i \cdot \bar{t}_i' = \rho_{ii}, \qquad \bar{t}_i' \cdot \bar{t}_j = \rho_{ij}$$

and so (5) is proved. For the other piece of information, from calculus, we recall how the vectors $\bar{t}_i$ were obtained; in particular, we recall that t_i depends only on the first i vector derivatives $\bar{x}'(s), \ldots, \bar{x}^{(i)}(s)$. So, the derivative $\bar{t}_i'$ of $\bar{t}_i$ depends only on the first $i + 1$ derivatives $\bar{x}'(s), \ldots, \bar{x}^{(i+1)}(s)$ or, what is the same thing, it depends only on $\bar{t}_1, \ldots, \bar{t}_{i+1}$. In view of (4) this means that

$$\rho_{ij} = 0 \qquad \text{if} \qquad j > i + 1. \tag{6}$$

Since the matrix R is also skew-symmetric, (6) means that R has a very special form: All elements vanish except those on the first superdiagonal and those on the first subdiagonal. Thus, with a slight change of notation for the coefficient ρ_{ij}, (4) becomes

$$\begin{aligned}
\bar{t}_1' &= \rho_1\bar{t}_2 \\
\bar{t}_2' &= -\rho_1\bar{t}_1 + \rho_2\bar{t}_3 \\
\bar{t}_3' &= -\rho_2\bar{t}_2 + \rho_3\bar{t}_4 \\
&\vdots \\
\bar{t}_{n-1}' &= -\rho_{n-2}\bar{t}_{n-2} + \rho_{n-1}\bar{t}_n \\
\bar{t}_n' &= -\rho_{n-1}\bar{t}_{n-1}.
\end{aligned} \tag{7}$$

These are the generalized Frenet-Serret equations, and the coefficients $\rho_1, \rho_2, \ldots, \rho_{n-1}$, which are of course functions of s, are the successive curvatures of K at the point p.

We have now obtained (7), by an algebraic-analytic sleight of hand, it might be said. It remains to show, for the case of three dimensions at least, that the formally obtained curvatures are indeed measures of curving and twisting. We shall do this next, restricting outselves to the case $n = 3$. First, in three dimensions there is a special time-honored terminology: the unit vectors $\bar{t}_1$, $\bar{t}_2$, $\bar{t}_3$ are relabeled as $\bar{t}$, $\bar{n}$, $\bar{b}$ and are called the tangent, the normal, and the binormal to K at p. Also, the quantities ρ_1 and ρ_2 are relabeled κ and τ and are called the curvature and the torsion of K at p. So with this new terminology the Frenet-Serret equations become

$$
\begin{aligned}
\bar{t}' &= & \kappa\bar{n} & \\
\bar{n}' &= -\kappa\bar{t} & & +\tau\bar{b} \\
\bar{b}' &= & -\tau\bar{n} &
\end{aligned}
\tag{8}
$$

A picture of the three vectors $\bar{t}$, $\bar{n}$, $\bar{b}$ at p is given in Fig. 10.1. The tangent line T is the first osculating flat F_1, the osculating plane O.S. through $\bar{t}$ and $\bar{n}$ is the second osculating flat F_2. We have previously given an informal description of curving as the tendency to depart from rectilinearity. Now the unit vector $\bar{t}$ is the direction of K at p and so it is natural to define curvature κ in terms of a derivative, i.e., the rate of change, of $\bar{t}$. Further, the arc length s is the natural parameter with respect to which to differentiate $\bar{t}$. This is so because s is an intrinsic quantity for the curve K and does not depend on the accidents of how K is located in space

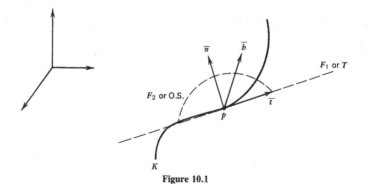

Figure 10.1

with respect to some coordinates. Thus $\bar{t}'$ is the measure of curvature; since it is a vector, its magnitude κ is the scalar quantity that measures curvature [compare with the first equation in (8)].

Next, twisting has been informally defined as the tendency of K to depart from planarity. Now the plane of best local fit to K at p is the osculating plane; we therefore want a quantity that measures how fast the osculating plane changes, or turns, as we go along the curve. The most natural vector to give the direction of a plane is the unit normal to the plane. In our setup such a unit normal to the osculating plane is the binormal vector $\bar{b}$. Hence the proper measure of twisting is the derivative $\bar{b}'$, and the proper scalar measure of twisting is the torsion τ, as we see from the last equation in (8). There is a slight difficulty with κ and τ, and this is the matter of sign; we can resolve it, e.g., by always taking $\kappa \geq 0$. Or κ can be positive or negative, by demanding that the sense of rotation from $\bar{t}$ to $\bar{n}$ should be the same as from the OX-axis to the OY-axis. Once the sense of $\bar{n}$ is chosen, the sense of $\bar{b}$ is fixed since $\bar{b} = \bar{t} \times \bar{n}$.

There is an alternative way of handling the curvature κ. We consider the circle C of best possible approximation to K at p. This circle is given more descriptively as the circle through three consecutive points of K at p (with the obvious interpretation of this phrase in the sense of limits). So C is the intersection of a sphere S whose center $\bar{c}$ lies in the osculating plane, and the osculating plane itself. If r is the radius of S then the equation of S is

$$(\bar{X} - \bar{c})\cdot(\bar{X} - \bar{c}) = r^2$$

where $\bar{X}$ is the general point of S. We need now the local condition on C to be the circle of best local fit to K at $p = \bar{x}(s)$. For this purpose we consider the scalar function

$$F(s) = (\bar{x} - \bar{c})\cdot(\bar{x} - \bar{c}) - r^2,$$

which measures the departure from points of C to points on K. For best local approximation the function F and its first two derivatives vanish:

$$F(s) = F'(s) = F''(s) = 0$$

or

$$(\bar{x} - \bar{c})\cdot(\bar{x} - \bar{c}) - r^2 = 0 \qquad (9)$$

$$(\bar{x} - \bar{c})\cdot\bar{x}' = 0$$

$$(\bar{x} - \bar{c})\cdot\bar{x}'' + 1 = 0.$$

The last equation is obtained because $\bar{x}'$ is a unit vector. Since $\bar{x} - \bar{c}$ lies in the osculating plane, which by definition is spanned by the vectors $\bar{x}'$, $\bar{x}''$, we have

$$\bar{x} - \bar{c} = c_1\bar{x}' + c_2\bar{x}'',$$

where c_1, c_2 are constants. It now follows from (9) that

$$(\bar{x} - \bar{c})\cdot\bar{x}' = c_1\bar{x}'\cdot\bar{x}' + c_2\bar{x}_2''\cdot\bar{x}' = 0$$

but $\bar{x}'\cdot\bar{x}' = 1$ and (hence) $\bar{x}''\cdot\bar{x}' = 0$ so that $c_1 = 0$. Further

$$c_2 = \frac{(\bar{x} - \bar{c})\cdot\bar{x}''}{\bar{x}''\cdot\bar{x}''} = \frac{-1}{\bar{x}''\cdot\bar{x}''}.$$

The first one of the Frenet-Serret equations (8) is $\bar{t}' = \kappa\bar{n}$ or $\bar{x}'' = \kappa\bar{n}$ (since $\bar{t} = \bar{x}'$), therefore

$$\bar{x}''\cdot\bar{x}'' = \kappa^2 \qquad\qquad (10)$$

and so, finally,

$$c_1 = 0, \qquad c_2 = \frac{-1}{\kappa^2}.$$

Therefore

$$\bar{c} = \bar{x} - c_2\bar{x}'' = \bar{x} - c_2\bar{t}' = \bar{x} + \frac{1}{\kappa}\bar{n};$$

giving to the reciprocal $1/\kappa$ its usual name of radius R of curvature we get

$$\bar{c} = \bar{x} + R\bar{n}.$$

This shows that the center of the osculating circle C lies on the principal normal, at the distance R from the point p of K (and of C). Thus the names circle of curvature and radius of curvature are justified. The formula (10) allows us to calculate κ and R provided that $\bar{x}$ is given in terms of the arc length s as parameter. If the vector $\bar{x}$ is given in terms of some other parameter u, it is not necessary to reparametrize to the arc length s; it can be shown that the curvature κ and the torsion τ are given as

$$\kappa^2 = \frac{(\dot{\bar{x}} \times \ddot{\bar{x}})\cdot(\dot{\bar{x}} \times \ddot{\bar{x}})}{(\dot{\bar{x}}\cdot\dot{\bar{x}})^3}, \tau = \frac{(\dot{\bar{x}}\ddot{\bar{x}}\dddot{\bar{x}})}{(\dot{\bar{x}} \times \ddot{\bar{x}})\cdot(\dot{\bar{x}} \times \ddot{\bar{x}})}$$

Here $\dot{\bar{x}}$ is $d\bar{x}/du$ and $(\bar{u}\,\bar{v}\,\bar{w})$ is the triple scalar product.

From the theory of differential equations it can be shown that if κ and τ are given as functions of s then the Frenet–Serret equations (8) determine $\bar{x} = \bar{x}(s)$ except for certain arbitrary constants. Thus the curve K itself is given completely by its curvature and torsion; the arbitrary constants control only the position of K in space. As a very special case let both the curvature and the torsion be constant: $\kappa = a$, $\tau = c$. Suppose first that $a = 0$, then from (8) $\bar{t}$ is a constant unit vector $\bar{t}_0$, and since $\bar{x}' = \bar{t}$ it follows that

$$\bar{x}(s) = s\bar{t}_0 + \bar{x}_0. \tag{11}$$

Thus K is a straight line in the direction $\bar{t}_0$ and passing through the point $\bar{x}_0$.

Next, let $a \neq 0$ but $c = 0$. We find now from (8) that $\bar{b}' = 0$ so that $\bar{b} = \bar{b}_0$: the binormal vector is constant. Hence the osculating plane P is constant and so K is a plane curve in P. The first two Frenet-Serret equations are

$$\bar{t}' = a\bar{n}, \qquad \bar{n}' = -a\bar{t};$$

differentiating the first equation and using the second one to eliminate $\bar{n}'$ we get

$$\bar{t}'' = -a^2\bar{t}. \tag{12}$$

This is the vector form of the simple harmonic equation and the solution is

$$\bar{t} = \bar{A}_0 \cos as + \bar{B}_0 \sin as$$

where $\bar{A}_0$, $\bar{B}_0$ are constant vectors. Since $\bar{t} = \bar{x}'$ we get

$$\bar{x} = -\frac{\bar{B}_0}{a} \cos as + \frac{\bar{A}_0}{a} \sin as + \bar{x}_0,$$

where $\bar{x}_0$ is a constant vector. We choose $\bar{x}_0 = 0$, which amounts merely to a suitable choice of the origin in the plane of K. Then

$$\bar{t} = \bar{A}_0 \cos as + \bar{B}_0 \sin as \tag{13}$$

$$\bar{x} = = \frac{\bar{B}_0}{a} \cos as + \frac{\bar{A}_0}{a} \sin as.$$

Since $\bar{t}$ is a unit vector we must have $\bar{t} \cdot \bar{t} = 1$ or

$$\bar{A}_0{}^2 \cos^2 as + 2\bar{A}_0 \cdot \bar{B}_0 \sin as \cos as + \bar{B}_0{}^2 \sin^2 as = 1$$

for all s. In particular, putting $s = 0$ and $s = \pi/2a$ we get

$$\bar{A}_0{}^2 = 1, \qquad \bar{B}_0{}^2 = 1 \tag{14}$$

and therefore also

$$\bar{A}_0 \cdot \bar{B}_0 = 0. \tag{15}$$

Now computing $\bar{t} \cdot \bar{x}$ from (13), and using (14) and (15), it is found that

$$\bar{t} \cdot \bar{x} = 0.$$

Therefore K is a circle since the radius vector $\bar{x}$ from the origin to any point on the curve is perpendicular to the tangent $\bar{t}$ at that point.

Suppose finally that $a \neq 0$ and $c \neq 0$. The Frenet-Serret equations are now

$$\bar{t}' = a\bar{n}, \qquad \bar{n}' = -a\bar{t} + c\bar{b}, \qquad \bar{b}' = -c\bar{n};$$

differentiating the middle one and using the others to eliminate $\bar{t}$ and $\bar{b}$, we get

$$\bar{n}'' = -m^2\bar{n},$$

where $m^2 = a^2 + c^2$. The preceding differential equation is solved just as (12) before was, and

$$\bar{n} = \bar{C}_0 \cos ms + \bar{D}_0 \sin ms. \tag{16}$$

Since $\bar{n}$ is also a unit vector, like $\bar{t}$ before, we find as before that

$$\bar{C}_0{}^2 = \bar{D}_0{}^2 = 1, \qquad \bar{C}_0 \cdot \bar{D}_0 = 0. \tag{17}$$

Since $\bar{t}' = a\bar{n}$ and $\bar{x}' = \bar{t}$, we have by integration

$$\bar{t} = -\frac{a\bar{D}_0}{m} \cos ms + \frac{a\bar{C}_0}{m} \sin ms + \bar{t}_0$$
$$\bar{x} = -\frac{a\bar{C}_0}{m^2} \cos ms - \frac{a\bar{D}_0}{m^2} \sin ms + \bar{t}_0 s + \bar{x}_0. \tag{18}$$

The constant vector $\bar{x}_0$ we take to be zero, since this only implies a change of the origin of coordinates. Also, $\bar{n} \cdot \bar{t} = 0$ since $\bar{n}$ and $\bar{t}$ are perpendicular. Computing $\bar{n} \cdot \bar{t}$ from (16) and (18), we find by using (17) that

$$\bar{t}_0 \cdot \bar{C}_0 \cos ms + \bar{t}_0 \cdot \bar{D}_0 \sin ms = 0$$

for all s. In particular, taking $s = 0$ and $s = \pi/2m$, we find that

$$\bar{t}_0 \cdot \bar{C}_0 = \bar{t}_0 \cdot \bar{D}_0 = 0. \tag{19}$$

Equations (17) and (19) show that every two of the three vectors $\bar{C}_0$, $\bar{D}_0$, $\bar{t}_0$ are perpendicular. Choose now a Cartesian coordinate system with the X-axis along $-\bar{C}_0$, the Y-axis along $-\bar{D}_0$, and the Z-axis along $\bar{t}_0$. Then the second equation in (18) gives the Cartesian equations of our curve K as

$$x = a_1 \cos ms, \qquad y = a_1 \sin ms, \qquad z = c_1 s;$$

therefore K is an ordinary circular helix H.

Our three self-congruent curves, the straight line L, the circle C, and the helix H, have thus been characterized in terms of curvature and torsion: L has zero curvature, C has constant curvature and zero torsion, H has constant curvature and constant torsion.

Our last topic is concerned with geodesics. These are curves of shortest length that join two points of a given surface S and lie themselves on S. Before taking up that topic it may be noted that the arcs of our three distinguished curves have the geodesic property. The shortest curve joining two points in the plane is a straight segment. For the sphere the shortest joining curve is, as we shall show, an arc of the great circle. The shortest-length curve joining two points on a circular cylinder Z is an arc of a circular helix. This last statement can be proved as follows: Unroll the cylinder surface Z onto a plane, and observe that the operation of unrolling *preserves lengths*. Therefore the shortest joining curve on Z unrolls onto a straight segment in the plane. Now execute the inverse operation: Roll the plane back onto the cylinder. It is a simple matter to show that the straight segment rolls up into an arc of a circular helix on Z.

The main difficulty with the shortest-distance definition of geodesics is that it does not characterize them by what we have called a local property, but only in the large, or globally. Now the geodesics on a surface S play the role of straight segments in the plane P. Also, as we know, straight lines and segments are characterized by the property of zero curvature. These considerations suggest that it might be possible to characterize the geodesics by some sort of a local curvature property. Indeed, there exists the so-called geodesic curvature, which concerns a local property of curves on a surface. Once this is introduced, it is then possible to define geodesics as curves on a surface having zero geodesic curvature. Such a definition is very convenient but too difficult to be used here.

We shall therefore use a theorem of the Swiss mathematician Johann Bernoulli (1667–1748) that provides a necessary condition characterizing

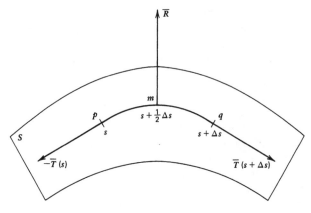

Figure 10.2

geodesics in simple and purely local terms:

> If G is a geodesic on the surface S, then at every point of G the principal normal to G coincides in direction with the surface normal to S.

The easiest way to demonstrate this is by a mechanical analogy: Stretch a thin elastic string, such as a rubber band, on the smooth surface S and find how the string lies on S. The physical condition of minimal potential energy characterizes the equilibrium position of the string, and this implies that its length is minimum. Hence the string lies along a geodesic curve on S. Consider now a small arc $\widehat{pq}$ of the string, of length Δs, as is shown in Fig. 10.2. Here s is as usual the arc length along the curve G of the string so that the point p corresponds to s and the point q to $s + \Delta s$. Three forces act on $\widehat{pq}$: the two tensions $-\bar{T}(s)$ and $\bar{T}(s + \Delta s)$ and the reaction $\bar{R}$ of S. The tensions are tangential to G and they may be expressed in terms of the unit tangent $\bar{t}(s)$ of G as $-k\bar{t}(s)$ and $k\bar{t}(s + \Delta s)$, where k is a constant. The reaction $\bar{R}$ is not localized at a point but is distributed all along the arc $\widehat{pq}$. So, instead of $\bar{R}$ we really have $\bar{r}\,\Delta s$, where $\bar{r} = \bar{r}(s)$ is the reaction per unit length. We locate the whole reaction $\bar{r}$ as at the midpoint m of $\widehat{pq}$, incurring thereby an error $\bar{E}\,\Delta s$ where $\bar{E}$ tends to zero together with Δs. Now the equation of equilibrium states that the net force on $\widehat{pq}$ vanishes:

$$-k\bar{t}(s) + k\bar{t}(s + \Delta s) + \bar{r}(s + \tfrac{1}{2}\Delta s)\,\Delta s + \bar{E}\,\Delta s = 0.$$

Dividing by Δs and letting $\Delta s \to 0$ we get in the limit

$$k\bar{t}'(s) + \bar{r}(s) = 0.$$

From the Frenet-Serret equations it follows that the derivative $\bar{t}'(s)$ is the principal normal $\bar{n}(s)$; since the surface S is assumed smooth, the reaction $\bar{r}(s)$ has the direction of the surface normal to S. This completes the proof.

Bernoulli's theorem will now be used to determine the geodesics on a sphere S. Let $\bar{x} = \bar{x}(s)$ be the vector equation of a geodesic G in a coordinate system whose origin is at the center of S. As usual, $\bar{t}$ and $\bar{n}$ denote the unit tangent and normal at a point of G. Also, let $\bar{N}$ be the unit normal to S so that $\bar{N} = (1/r)\,\bar{x}$, where r is the radius of S. By Bernoulli's theorem

$$\bar{N} = \bar{n}$$

so that

$$\bar{n}' = \frac{1}{r}\bar{t}$$

and therefore by the Frenet-Serret equations

$$\bar{n}' = -\kappa\bar{t} + \tau\bar{b} = \frac{1}{r}\bar{t}.$$

It follows that $\tau = 0$ so that G is a plane curve on S, it is therefore an arc of a circle. But, further, the curvature κ of G is $1/r$, where r is the radius of the sphere S. Hence G is an arc of a great circle on S.

EXERCISES

1. Show that the vector equation of the complete space curve which is the intersection of the sphere $x^2 + y^2 + z^2 = 4a^2$ with the cylinder $(x - a)^2 + y^2 = a^2$, is

    $$\bar{x} = [a(1 + \cos 2t),\ a \sin 2t,\ 2a \sin t], \qquad 0 \le t \le 2\pi.$$

2. Let G be a geodesic on a polyhedron P. How does G cross an edge E of P? (*Hint*: Unroll onto a plane and show that the two straight segments of G make equal angles with E.)

3. Show that on a regular tetrahedron T there is a plane closed geodesic G forming a square. Show that any parallel geodesic is also closed

and has the same length as G. What is its shape? (*Hint*: Use a suitable plane net of T.)

4. Let G be any geodesic on a circular cone C and p a point on G. Let r be the distance of p from the axis of C and let ϕ be the angle at which G cuts the generator of C through p. Show that $r \sin \phi$ is constant as p varies on C. (*Hint*: Unroll C onto a plane.)

5. Let a polycone be the surface of revolution arising by rotating a plane piecewise linear arc C about a line L in the plane of C, which does not cross C. Let G be a geodesic on a polycone. Show that $r \sin \phi$ is constant along G. (*Hint*: Use the preceding problem and show that $r \sin \phi$ does not change when we cross from one conical component of the polycone to another.)

6. Let S be any surface of revolution and G a geodesic on it. Let r be as in problem 4 and let ϕ be the angle at which G cuts the meridian of S (that is, a plane section of S by a plane through the axis of S). Prove the theorem of Clairaut that along G $r\sin\phi$ is constant. (*Hint*: Approximate to S by a polycone and use problem 5.)

7. A general helix is a space curve whose unit tangent makes a constant angle with a fixed line in space. Show that for a general helix the ratio of curvature to torsion is constant. (*Hint*: Use the Frenet-Serret equations.)

8. Formulate and prove the statement converse to the last one.

9. Show that the general helix on the cone $x^2 + y^2 = a^2 z^2$ projects onto the xy-plane as a logarithmic spiral.

10. Use Bernoulli's theorem to show that a geodesic on a circular cylinder is an arc of a circular helix.

NOTES

CHAPTER 1

1. If the beginning is too sudden, it is possible to produce a more gradual introduction. Probably the best way is to summarize the work of Thales on the proportionality of triangles and his measurement of the height of a pyramid in Egypt: He observed that by similarity of triangles the height of a pyramid is to the length of its shadow as the height of a vertical rod to the length of *its* shadow. This may be followed by the theorem of Thales proved in the text, that all angles in a circle based on the same chord are equal; Thales also observed that all angles on the diameter are right.

2. There are other proofs, or almost-proofs, of the isoperimetric property of the circle (the "almost" part refers to lack of existence proof—that a maximizing curve exists). One of the simplest, due to Steiner, goes as follows. First, the closed plane curve C of fixed circumference must be convex if it is to enclose maximum area; otherwise the reentrant parts could be reflected outward in their double tangents. Next, a line L that cuts C into two parts of equal length also cuts the area enclosed by C into two equal parts. For otherwise, we could replace C by one such part, together with its reflection in L. Let L cut C in points A and B, and let X be any other point on C. Consider now that part of the area of C that lies on the same side of L as X. It consists of three parts: the triangle AXB, and the two others bounded by AX, BX, and by the arcs of C. The angle of ABX at X must be 90° since otherwise the half of C can be perturbed without changing its length so that the total area increases ("slide" A and B on L toward each other or away from each other, so that the angle at X becomes 90° while the two parts bounded by the arcs keep their exact size and shape). But X is arbitrary; therefore C is a circle with diameter AB.

So, why do our proof, which is considerably longer? In the first place, we get the Heron and Brahmagupta formulas, and other useful side products. But more importantly, our proof is of a much more general type whereas Steiner's proof is completely special. In fact, our proof forms a small introduction to the Ritz method for direct solution of a variational isoperimetric problem. Here a certain quantity (for us, area) associated with an unknown curve is to be extremized while another such quantity (for us, length) is kept fixed. The Ritz method depends on a discretization: Approximate to the unknown curve, for instance, by a piecewise linear curve, and find the best approximation by ordinary calculus, or otherwise. This is exactly what we do with the polygonal approximation to the circle.

3. Why go into the existence of cyclic polygons with prescribed sides? First, a neat mechanical-analogy method is displayed. Also, area calculations can now be carried out. But perhaps the most important reason is to show that something we have had for more than 2000 years can be generalized and something new can be done with it. We speak here of something as simple as the ancient division of triangles into acute, right, and obtuse. Although every triangle is automatically cyclic, not every convex n-gon is so ($n > 3$). If it is, then its circumcenter is either inside it, on it, or outside, etc.

4. Easy formulas for the area of the triangle and of the cyclic quadrilateral are derived in the text, in terms of their sides. A method is outlined that allows us in principle to do the same for the general cyclic n-gon. But is there an explicit formula for the area? In analogy to the work of Abel and Galois, it may be conjectured that no explicit algebraic area formula exists for the general cyclic n-gon, if $n \geq 5$.

CHAPTER 2

The main feature of this chapter is the great compression of its material. A single two-part theorem of Routh is proved; then the important results of Ceva and Menelaus follow from it. It is worth emphasizing here that the pure geometry theorems of Ceva and Menelaus follow from the more general result of Routh, which he has obtained in connection with an applied mechanics problem involving stress and strain analysis in frameworks. The theorems of Ceva and Menelaus give convenient criteria for three lines to be concurrent and for three points to be collinear. In particular, it follows that in any triangle the three angle bisectors pass through one point, and so do the three medians, the three heights, etc. Thus the existence of the in-center, centroid, orthocenter, etc., is shown by a uniform method. The ω-point construction, for $\omega = 60°$, will be of importance later when we come to the Steiner problem in Chapter 8, on how to connect n points by the shortest possible network.

The material at the beginning of the section allows, if desired, an easy introduction to harmonic quadruples: If A, X, B, Y are four points on a line, then the quadruple of points is called harmonic if X divides AB internally in the same numerical ratio (say, r) as Y does externally (say, $-r$). One use of this is made in the ruler-alone construction illustrated in Fig. 2.3a and b.

The extension of the Euclidean to the projective plane can be handled by easy stages that emphasize the convenience of the "fictional" points and "fictional" line, at infinity. The convenience consists, for us, in avoiding special cases, as, for instance, in Pascal's theorem of Chapter 5.

CHAPTER 3

1. The interest in the geometry of cycloids and epicycloids is very old; it certainly precedes the discovery that the cycloid has both the brachistochrone and the tautochrone properties. We may connect that interest with the pre-Copernican astronomical system called Ptolemaic (named after the same Alexandrian Greek who discovered "Ptolemy's theorem" of Chapter 1). The planetary paths were first assumed to be circles; then as more exact astronomical observations became available, it was found that simple circular orbits did not fit the data. On account of some philosophical penchant for the perfection of circles, the necessary corrections were attempted by putting into the system circles that rolled on and within other circles, the so-called epicycles. Eventually the system broke down and epicyclic orbits were replaced by Kepler with elliptical ones.

2. We show that a particular hypocycloid reduces to a straight segment. Here it is possible to insert additional material on the Peaucellier and Hart linkages, and on other exact geometrical methods of generating a straight line.

3. Two remarks may be added to our generalized corridor problem of the longest straight rod that can be moved through a junction of corridors of widths a and b, meeting at the angle α. First, if α is

small enough, we ought to consider a different type of minimum in which the rod, instead of being turned round the corner, is moved into the opposite corner, turned through the angle α, and then moved along the other corridor with its ends reversed (that is, it moves first as AB, then as BA). Second, there is a great difference of principle between classical Euclidean geometry, which insists on exact constructions, and the solution given here, which is more in the spirit of numerical analysis (where the object is to produce a sufficiently, or arbitrarily, good approximation, rather than an exact answer).

4. The Descartes principle, that a plane rigid motion is at each instant a rotation about some instantaneous center, may be interpreted as showing Descartes to be one of Newton's predecessors in the discovery of calculus. To some extent, the same is true of Pascal with respect to the role of Pascal's theorem in enabling us to draw tangents to conics.

CHAPTER 4

1. The material on the symmetries of the cube can be developed into a geometrical introduction to the elements of group theory. For instance, one may compute symmetries inverse to a given one, show that symmetries A and B exist such that AB is different from BA, and so on, up to setting up a complete multiplication table for all symmetries of the cube. In a similar way, the material concerning the winding types of closed curves on the torus that pass through a fixed point can be augmented to provide a simple introduction to what is known in topology as homotopy with base point.

2. The work with plane sections of a torus shows something mildly surprising: Any such section has an explicit equation of the form $y = f(x)$ involving square roots only and nothing more complicated than that. However, there may be a square root under a square root so that in plotting the section curves one must consider only those values of the variable that do not lead to square roots of negative quantities. The plotting of several different plane sections might be a good, if slightly strenuous, exercise. In the first place, students nowadays rarely get acquainted with any plane curves beyond the simplest ones. But further, our plane sections provide a good, though very simple, introduction to what may be called geometrical bifur-

cation. The phenomenon of bifurcation concerns the qualitative change of trajectories in differential systems near critical points and would usually be studied only in an advanced course on differential equations. To see a type of bifurcation, one need only consider a horizontal torus and its plane section by a moving vertical plane, as in our Exercise 8. At the critical place the plane section has the shape of a figure eight; as the plane moves, in one direction we get a single closed curve with a narrow isthmus, and in the other direction the figure eight separates into two distinct closed curves.

3. With access to a computer and with some work in computer graphics, the sections of the cube and the torus may be displayed on a screen. Further, different sections might be displayed by controlling the parameters that fix the cutting plane.

CHAPTER 5

1. Like Chapter 1, this one is of central importance, and for several reasons: historical, methodological, (purely) geometrical, applied (e.g., to physics and astronomy), and so on. This may be seen already from the span of time covered (from ancient Greece to modern France) and from the variety of subjects touched on (from the elements of differential equations to a very pure type of geometry as in the theorems of Pappus and Pascal). It is sometimes claimed that the Greek achievement of a reasonably complete theory of conic sections is one of the finest achievements of ancient science. This chapter gives some of the background by outlining the Greek synthetic method; we also contrast this with the modern analytic method, which handles geometry by means of equations.

2. The Greek geometry arose partly in the attempt to solve the three classical problems: to trisect an angle, to duplicate the cube, to square the circle; this is especially true of the geometry of conic sections. The ancient Greeks allowed themselves only a very limited freedom of geometrical construction: the use of ruler and compasses alone. But one must be careful here, and the Greeks themselves realized that a seemingly small departure in employing the ruler will make a difference in what problems can be solved. To show this, we give the very simple method of trisecting any angle, due to

Archimedes. This method uses the ruler and compasses, but one is allowed to make marks on the ruler's edge (two marks are enough).

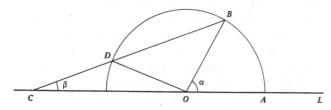

By repeated bisection, if necessary, it may be assumed that the angle α to be trisected is less than the right angle. Let O be a point on the straight line L, draw the semicircle centered at O of some radius OA, and let B be the point such that $\measuredangle AOB = \alpha$. On the straight ruler mark two points C and D such that $CD = OA$. Now slide the ruler so that C moves on L while D slides on the semicircle. At some place the edge of the ruler will pass through B; this is the position shown in the figure. Then the triangle CDO is isosceles (because $CD = OA = OD$) so that $\measuredangle OCD = \measuredangle COD = \beta$, say. Therefore $\measuredangle BDO = 2\beta$ and hence $\measuredangle OBD = 2\beta$ also. But $\measuredangle DCO + \measuredangle DBO = \measuredangle BOA$ so that $3\beta = \alpha$; α has been trisected.

3. Newton's role in the foundation of calculus is well known; he used the machinery of calculus to show how the laws of Kepler can be derived from the laws of motion and of universal gravitational attraction. Thus he showed that under suitable assumptions the orbits of planets and comets are indeed conic sections. Less well known is the fact that Newton worked in geometry as well. Among other things, he proved an odd property of the ellipse: If an ellipse is inscribed into a convex quadrilateral, so that it touches all four sides, then its center must lie on the straight segment joining the midpoints of the quadrilateral's diagonals. Now, the conic sections may be called quadratic curves; they are given by equations of the form $P(x, y) = 0$, where P is a quadratic polynomial. If P is allowed to be a cubic polynomial instead, then we get the so-called cubic curves. On a much more fundamental level, Newton obtained a classification of cubic curves; this is analogous to classifying quadratic curves into ellipses, parabolas, hyperbolas, and the degenerate cases. It may be that Newton's interest in cubic curves was stimulated by his discovery that conic-section orbits follow from the inverse-square law of force, and by a desire to investigate what shape the orbits might

have under laws of attractive force other than the inverse-square case.

4. The method used here to prove Pascal's theorem depends on taking linear combinations: Let $F_1 = 0$ and $F_2 = 0$ be the equations of two loci of the same type, for instance, lines, circles, or conics; then $F_1 + kF_2 = 0$ is the equation of a like locus that passes through the intersection of the two original ones. The same applies in space rather than the plane, with respect to planes, spheres, and so on. This method is of general usefulness in elementary, as well as less elementary, analytic geometry. For instance, it enables us to solve very simply such "intersection-type" problems as the following: finding the straight line (or the plane) that passes through the intersection of two given lines (or planes), and through a given point (or has a prescribed slope); finding the straight line (or plane) that passes through the intersection of two intersecting circles (or spheres); or finding the conic section that passes through the four points in which two given lines cut a given conic section, and through one other point.

An interesting possibility arises when the curves (or surfaces) given by $F_1 = 0$ and $F_2 = 0$ do not intersect. What is $F_1 + kF_2 = 0$ then? Let the curves be two circles external to each other; then $F_1 - F_2 = 0$ is the equation of a straight line, the so-called radical axis of the two circles. It has the property of being the locus of centers of all circles that cut the given two orthogonally. With this as the basis it is not hard to solve the following problem: Given three pairwise external circles, find the (unique) circle that cuts all three orthogonally.

5. A brief comment must be made on the theorem of Pappus, from this chapter, and the theorem of Desargues, from Chapter 2. In the usual elementary treatment of analytic geometry we start with a certain special algebraic domain (in fact, the field), namely, the real numbers, and we use its members, the real numbers, to coordinatize points with; then we derive equations of lines, curves, and so on. In this process one uses the standard algebraic laws obeyed by real numbers: the associative law [$x(yz) = (xy)z$ for all real x, y, z], the commutative law ($xy = yx$ for all real x, y), and others.

In an advanced treatment it is possible to reverse this procedure to some extent. One starts with an abstract "geometry" and introduces in it such geometrical entities as points and lines, by means of axioms. Then it is possible to ask, Are the points coordinatized by assigning to them members of some algebraic domain (or pairs, or triples of such members)? That is, one asks, What *algebraic* properties of the coordinatizing domain follow from the purely *geometrical* axioms assumed to hold? It is known from the work of D. Hilbert on the foundations of geometry that if the theorem of Desargues is one of the axioms of such abstract geometry, then the algebraic coordinate domain satisfies the associative law. In the same way, the theorem of Pappus implies the commutative law.

6. The final remark concerns the theorem of Brianchon, which is dual to that of Pascal. We start by observing that in the projective plane there is a duality between points and lines: A valid proposition about points and lines remains valid if "point" and "line" are interchanged (allowing for some slight vagaries of language). Simple examples of such dual pairs are as follows: (1) on any line there are infinitely many points—through any point there are infinitely many lines, or (2) two distinct points determine a unique line—two distinct lines determine a unique point. In the Euclidean plane this is not so; two distinct parallel lines do not determine any point. Such projective duality produces new theorems out of old ones. If it is applied to Pascal's theorem, then there results the dual theorem of Brianchon: If a hexagon is circumscribed about a conic (so that its sides touch it), then the three lines joining the opposite vertices intersect in a point.

CHAPTER 6

1. The eight problems on geometrical extrema collected here are very unequal in importance. The first few supply merely a convenient hook on which to hang something; the last two provide introductions to whole new geometrical topics. In particular, Problem 7 leads to sets of constant width, and Problem 8 introduces the subject of packings and coverings. However, even though the first few problems are in themselves of no particular importance, the techniques used to solve them are important; this is especially true about the reflection technique employed in Problems 4 and 5.

2. The reflection method used in Problems 4 and 5 introduces the possibility of a new type of geometrical construction that uses neither ruler nor compasses: paper folding and pin. Consider, for instance,

the problem illustrated in Fig. 6.4*b* of producing the shortest path *APQB*, given the points *A* and *B*, and the lines L_1 and L_2. By folding the page flat so that the straight crease is L_1, and using the pin, we get the point A_1. Then, unfolding and producing another fold whose crease is L_2, we get A_2. Now the points *Q* and *P* can be found. Similar paper folding will produce the orthic triangle which is the solution of Fagnano's Problem 5. The paper-folding-and-pin method is not entirely trivial; this may be seen by asking the following question: Does it, or does it not, enable us to find the Steiner point *S*, given points *A*, *B*, *C* as in Fig. 6.7?

3. Consider the acute-angled triangle *ABC* of Fig. 6.5*c* as a flat and very thin triangular parcel. Then the closed six-segment path $C_2A_2B_2C_3A_3B_3C_2$ may be regarded as a closed loop of thin inextensible string that is stretched taut and ties the parcel up. The six straight segments lie alternately on the top face and on the bottom face. In more formal language we might say that the loop is a closed geodesic on the thin triangular parcel. Can the string be taken off the parcel, without stretching or cutting? The answer is yes. Since all such loops have the *same length*, we can deform the string to another configuration in which each of the six straight portions moves into a new position, parallel to the old one. Deforming it thus, we eventually get the string to, and over, a vertex of the triangle,

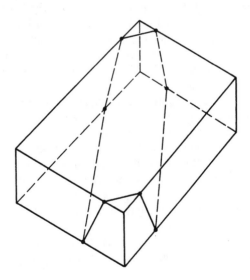

and now the string comes off. The same question may be asked for a stretched string that ties up a rectangular box (not necessarily "thin") as shown above. Here we have eight straight segments, and everything else is as before. The answer is again yes, and the sole difficulty is in showing that the total length for a parallel position is the same. This is simply accomplished by using a plane development of the faces of our box, of the staircase type, illustrated in Fig. 4.4a. In that plane face network two of the six rectangular faces are repeated twice; thus there are eight rectangles corresponding to the eight straight portions.

4. The physical-analog method for solving the Steiner problem is illustrated in Fig. 6.6a. Here W_1, W_2, W_3 are three equal weights, and in the equilibrium position the point X where the three threads meet comes to rest at the Steiner point for the triangle ABC. Thus the total length $AX + BX + CX$ is minimized. It may be of interest to investigate the case of unequal weights, or of uniform heavy threads. This is with special reference to minimizing a weighted sum $aAX + bBX + cCX$, where a, b, c are three given positive numbers.

5. The sets of constant width introduced in Problem 7 have a rich structure and have commanded the interest of mathematicians since Euler. There are several equivalent characterizations besides the defining one that the width is the same in every direction. For instance, given a bounded set X in a Euclidean space (of any dimension), let its diameter be defined as the supremum of the lengths of all straight segments xy, where x and y are points of X. It turns out that X is of constant width if and only if on adding to X any point not already in X the diameter is increased.

   It may be noted that there is no equivalent of Reuleaux polygons in three dimensions. For instance, take a regular tetrahedron T and let R be the region common to the four balls each of which is centered at a vertex of T and passes through the other three vertices. It might be thought that R is of constant width; however, it is not. Consider the six circular arcs that are the edges of R; it turns out that any two opposite ones (i.e., any pair of edges without a common point) have midpoints that are slightly further apart than any two vertices of R.

   Of the many theorems for curves of constant width we quote one, due to a French mathematician Barbier: Any curve C of constant width D has the circumference πD. A sketch of the proof may be based on the idea of approximating arbitrarily well to C by a Reuleaux polygon P with sufficiently many vertices, and of the same width D as C itself. Let any double normal in P be considered and rotate it, each time about a vertex of P, from some initial position A_1A_2 to the final position A_2A_1. Then the total length of P is D times the angle through which the double normal has been turned, but that angle is π.

6. Problem 8 barely scratches the surface of the very rich subject of packing and covering. The problems of packing are of considerable interest in other branches of mathematics, for instance, in number theory, and even outside mathematics, for instance in communications theory (in connection with the so-called theory of error-correcting codes). The best geometrical introduction is the book of Fejes-Toth, referred to in the bibliography. We mention here the problem of the thinnest covering, which is in a sense dual to our Problem 8 of the densest packing. What is the thinnest covering of the whole plane by overlapping unit circles? Here the packing fraction exceeds 1 since the circles must overlap in order to cover the whole plane. It turns out that, for the thinnest covering, the centers of the circles form again a regular hexagonal array, and the lowest density is $2\pi/\sqrt{27} = 1.209. \ldots$

CHAPTER 7

1. As in Chapters 1 and 5, the historical element is strong here. We have the old Egyptian empirical observation that a hemisphereical basket needs twice as much material as its flat circular cover; then comes the fundamental work of Archimedes concerning the sphere and the cylinder; the generalized theorem of Archimedes is shown to have relevance to a modern problem in probability (the random walk in three dimensions). The spherical-excess formula for the area of a spherical triangle is due to the French mathematician A. M. Legendre (1752–1833). It is a very special case of a deep generalization, known as the Gauss-Bonnet theorem, which is one of the starting points of the differential geometry of surfaces.

2. Steradians and steradian content are of some importance in lighting engineering and, even more, in radio engineering. The questions of

interest here concern the density of light, or other form of radiation, projected in a beam. Such a beam often has a conical shape, the cone being a general one and not necessarily circular, projected from an optical reflector or from an antenna.

3. The following remarks explain some of the conceptual background behind the generalized theorem of Archimedes. It may be said that this theorem depends on transferring a certain simple but useful principle from the discrete domain of counting to the continuous domain of measuring. The discrete counting principle is, roughly, this: Two finite sets are equinumerous if a one-to-one correspondence exists between their members. Thus there is no need for any counting if a large number of cups is present on a table and each cup stands on its own saucer while every saucer has a cup on it; we know that the number of cups equals the number of saucers. In spite of its simplicity the principle is heavily exploited in the subject of combinatorics, mainly to reduce harder counting problems to easier ones, but also in other ways. A somewhat spectacular application depends on using this principle in the negative: Two sets cannot be equinumerous if there is no one-to-one correspondence between their members. The application is as follows. An ordinary 8-by-8 chessboard is covered by 32 dominoes, each of which is a rectangle covering exactly two squares of the chessboard. Next the chessboard is mutilated: some two squares on a diagonal line are removed. Can the mutilated board be covered exactly by 31 dominoes? The answer is no, but a direct attempt to prove it is almost certain to fail. However, our counting principle can be invoked. Each domino must cover two neighboring squares of the board, and one of these is black, the other white. Thus, if the covering were possible, there would have been as many white squares covered as black ones. But the two removed squares are necessarily of the same color being on one diagonal line. Hence the 62 squares of the mutilated board have 30 of one color and 32 of the other, and so no one-to-one correspondence is possible between the black squares and the white ones; therefore no covering is possible.

The foregoing may be granted to be useful and/or amusing, but is there any connection with the generalized theorem of Archimedes? Certainly. Instead of counting we have now measuring (of areas) and instead of discrete members making up a set we have now certain "elementary parts." More exactly, since areas are computed by means of integrals, the "elementary parts" are the area elements used in integration. The role of the one-to-one correspondence is now taken over by the axial projection and so the area elements on the cylinder and on the sphere are equal. Hence the whole areas of the two corresponding sets, the one on the cylinder and the other on the sphere, are also equal.

There is another theorem, also concerning measurement though of volume rather than of area, that obviously has the same conceptual background. This is the theorem of B. Cavalieri (1598–1647), antedating the discovery of calculus. Its approximate sense is that two different solids are equal in volume if they are composed of the same "volume elements"; the theorem itself is as follows. Consider two solids in three dimensions that can be moved to positions X and Y in which there is a line L such that every plane perpendicular to L cuts X and Y in sections of equal area; then the two solids have equal volumes. Cavalieri's theorem is useful in deriving certain formulas for volume, for instance, the so-called prismoidal formula. We describe its somewhat complicated background and give the formula itself, since it applies to the volumes of cylinders, pyramids, and cones (both truncated and untruncated), certain slices of one-sheeted hyperboloids, and many other cases. In itself the prismoidal formula is simply proved by combining Cavalieri's theorem with an elementary application of the Simpson rule. First, a prismoid P is defined. This is any bounded solid having a flat top T and a flat base B; these two lie in parallel planes, but are allowed to shrink to a point or a segment. Further, the lateral surface bounding P must consist of complete straight segments joining points on the rim of T to points on the rim of B. However, there may be a "hole" through P, provided that it runs from T to B and is similarly generated by complete straight segments. Let h be the height of P, that is, the distance between the planes containing T and B. Let M be the cross section of P by the plane halfway between T and B. Then the prismoidal formula is

$$\text{vol } P = \frac{h}{6} [\text{area } T + 4 \text{ area } M + \text{ area } B].$$

4. Once upon a time, when the spherical trigonometry was still a popular subject, the proofs of the spherical laws of sines and cosines were given by geometry. Now that geometry has fallen into desue-

218 INVITATION TO GEOMETRY

tude and spherical trigonometry is nearly forgotten, the usual proofs
of those laws start from the well-known vector identities

$$(\bar{a} \times \bar{b}) \cdot (\bar{c} \times \bar{d}) = (\bar{a} \cdot \bar{c})(\bar{b} \cdot \bar{d}) - (\bar{a} \cdot \bar{d})(\bar{b} \cdot \bar{c}),$$

$$(\bar{a} \times \bar{b}) \times (\bar{c} \times \bar{d}) = (\bar{a}\bar{b}\bar{d})\bar{c} - (\bar{a}\bar{b}\bar{c})\bar{d},$$

where $(\bar{a}\bar{b}\bar{c})$ is the usual triple product $(\bar{a} \times \bar{b}) \cdot \bar{c}$. In fact, such
proofs, together with other snippets of spherical-trigonometric lore,
are inserted into many texts on vector algebra. Since students today
are quite likely to be reasonably acquainted with vectors and their
use but rather ignorant of geometry, I have preferred to give geometrical proofs rather than use vectors.

CHAPTER 8

1. Graph theory is today a thriving separate branch of mathematics
   with close relations to combinatorics, topology, algebra, logic and
   foundations, etc. It has numerous applications inside mathematics,
   both pure and applied, and outside it—for instance, to a branch of
   physics called statistical mechanics. Here we merely scratch the
   surface of the subject by collecting together a few funadmentals of
   graph theory that appear to be more geometrical in nature. One of
   our main objects is to develop sufficient machinery to be able to
   handle Steiner's problem which occupies almost half of this section.

2. The problem of the seven bridges of Königsberg, solved by Euler
   in 1736, is sometimes mentioned as one of the early starting points
   of topology. It certainly shows in a relatively simple way a very
   important aspect of topology, that it is a study of the connectivity
   properties of geometrical structures.

3. In the Theseus problem of traversing the labyrinth and finding the
   Minotaur, it is obvious that the search may be considerably simpli-
   fied. In particular, there is no need to return each time to the entrance. For instance, with reference to Fig. 8.4, suppose that Theseus has executed the two trips of length 1 without finding M. He
   then moves along corridor a to vertex B, and from B he executes
   all trips of length 1 (corridors h, g, f, e, d, c) returning each time
   to B only, not to A. Since M is not found, he returns to A, goes along
   b to C, and from C as temporary base he examines corridors c, i,

j, *k*, returning each time to *C*. Again, no *M* is found. So he returns to *A* and starts along *a*, then along *h*, to reach vertex *H*, and repeats the procedure with *H* as temporary base, etc.

CHAPTER 9

Convexity is a very extensive subject with many ramifications; its applications range from the purest number theory to the most applied optimization techniques. We have already come across convexity ideas in Chapter 1, in connection with the isoperimetry of polygons of prescribed side lengths, and in Chapter 6, in connection with Voronoi polygons. The present chapter has two main objectives: (a) to introduce in an easy and informal way some of the geometrical fundamentals of convexity in two and three dimensions, and (b) to provide a convenient place for such elements of the topology of point sets as open and closed sets, connected sets, boundary and interior points, etc. Several quite different definitions of convexity are given. It would be a good exercise to arrange them in a suitable circular order and to show that each implies the next all the way round the circle. This would constitute a formal proof of the equivalence of our definitions.

CHAPTER 10

The subject of differential geometry, whether in its classical or modern treatment, is perhaps one of the most important ones in the whole of mathematics. However, it is so rich and extensive, has such a variety of applications, and uses so many advanced mathematical techniques from algebra, topology, and analysis that not even an introduction could be given at our elementary level. Still, it is such an important branch of geometry that some rudimentary treatment seemed necessary in this book; it was only a question of what topic to choose. The selection made was motivated by three considerations: (1) importance, (2) simplicity, (3) illustrative power. The first point is obvious: The concepts of curving and twisting are surely among the most fundamental ones in differential geometry. As to point 2, our treatment of curvature and torsion is tied to the striking self-congruence property of the line, the circle, and the circular helix. A simple mechanical picture shows how these three curves can "ride on themselves"; this picture serves as a basis for our discussion of curving and twisting. As to point 3, we mention two different things. First, there is the interaction of different aspects of mathematics. Even the very name "differential geometry" implies such an interaction: that of tools from analysis and concepts from geometry; additionally, algebraic and topological techniques would also come into the subject. Now, at our very modest level, we illustrate such an interaction. The Frenet-Serret formulas are obtained by applying to a *geometrical* situation certain simple tools from *analysis* and from *linear algebra*. The second aspect of point 3 concerns one of the most basic techniques in the whole of mathematics: that of approximating locally to something complicated by something simple. The simplest example is of course met in calculus: The tangent line is the best local linear approximation to a curve, and the tangent plane to a surface. If the best linear fit is not good enough and better approximations are needed, one may use the Taylor polynomials, that is, longer initial segments of the Taylor series. In a different approach the circle (or the sphere) of the tightest local fit may be introduced, leading to the concept of curvature and radius of curvature. This technique of best local fit is called the approximation by osculating structures; the word *osculating* means literally "kissing." In our treatment the space curve C in n-dimensional space is locally approximated by its osculating k-flats, and as a result we obtain the curvature description of C.

BIBLIOGRAPHY

Artobolevskii I. I., *Mechanisms for the Generation of Plane Curves*. New York: Macmillan, 1964.

Berge, C., *The Theory of Graphs*. New York: Wiley, 1962.

Bonnesen T. and Fenchel W., *Theorie der konvexen Körper*. New York: Chelsea, 1948.

Boyer C. B., *A History of Mathematics*. New York: Wiley, 1968.

Briot C. A. and Bouquet J., *Analytical Geometry of Two Dimensions*, 14th ed. Chicago, 1896.

Cantor M., *Vorlesungen über Geschichte der Mathematik*, 4 vols. Boulder, CO: Johnson Reprint, 1965.

Coxeter H. S. M., *Introduction to Geometry*. New York: Wiley, 1961.

Dörrie H., *Hundred Great Problems of Elementary Mathematics*, 2nd ed. New York: Dover, 1965.

Eves, H., *A Survey of Geometry*, rev. ed. Boston: Allyn and Bacon, 1972.

Fejes-Toth L., *Lagerungen in der Ebene, auf der Kugel, und im Raum*. Berlin: Springer, 1953.

Hilbert D. and Cohn-Vossen S., *Geometry and the Imagination*. New York: Chelsea, 1952.

Hobson E. W., *Squaring the Circle and Other Monographs*. New York: Chelsea, 1950.

Lyusternik L. A., *Shortest Paths*. Elmsford, N.Y.: Pergamon, 1964.

McClelland W. J. and Preston J., *A Treatise on Spherical Trigonometry*. New York: Macmillan, 1912.

Meschkowski H., *Ungelöste und unlösbare Probleme der Geometrie*. Braunschweig: F. Vieweg & Sohn, 1960.

Smith D. E., *The Teaching of Geometry*. Boston: Ginn, 1911.

Steinhaus H., *Mathematical Snapshots*, 3rd Amer. ed. New York: Oxford University Press, 1969.

Struik D. J., *Lectures on Classical Differential Geometry*. Boston: Addison-Wesley, 1950.

Yaglom I. N. and Boltyanskii V. G., *Convex Figures*. New York: Dover, 1961.

INDEX

223